IEA Research for Education

A Series of In-depth Analyses Based on Data of the International Association for the Evaluation of Educational Achievement (IEA)

Volume 9

The International Association for the Evaluation of Educational Achievement (IEA) is an independent nongovernmental nonprofit cooperative of national research institutions and governmental research agencies that originated in Hamburg, Germany in 1958. For over 60 years, IEA has developed and conducted high-quality, large-scale comparative studies in education to support countries' efforts to engage in national strategies for educational monitoring and improvement.

IEA continues to promote capacity building and knowledge sharing to foster innovation and quality in education, proudly uniting more than 60 member institutions, with studies conducted in more than 100 countries worldwide.

IEA's comprehensive data provide an unparalleled longitudinal resource for researchers, and this series of in-depth peer-reviewed thematic reports can be used to shed light on critical questions concerning educational policies and educational research. The goal is to encourage international dialogue focusing on policy matters and technical evaluation procedures. The resulting debate integrates powerful conceptual frameworks, comprehensive datasets and rigorous analysis, thus enhancing understanding of diverse education systems worldwide.

More information about this series at http://www.springer.com/series/14293

Teresa Neidorf · Alka Arora · Ebru Erberber ·
Yemurai Tsokodayi · Thanh Mai

Student Misconceptions and Errors in Physics and Mathematics

Exploring Data from TIMSS and TIMSS Advanced

IEA

Springer Open

Teresa Neidorf
American Institutes for Research
Washington, DC, USA

Alka Arora
American Institutes for Research
Washington, DC, USA

Ebru Erberber
American Institutes for Research
Washington, DC, USA

Yemurai Tsokodayi
American Institutes for Research
Washington, DC, USA

Thanh Mai
American Institutes for Research
Washington, DC, USA

ISSN 2366-1631 ISSN 2366-164X (electronic)
IEA Research for Education
ISBN 978-3-030-30187-3 ISBN 978-3-030-30188-0 (eBook)
https://doi.org/10.1007/978-3-030-30188-0

This Springer imprint is published by the registered company Springer Nature Switzerland AG
The registered company address is: Gewerbestrasse 11, 6330 Cham, Switzerland

Foreword

IEA's mission is to enhance knowledge about education systems worldwide and to provide high-quality data that will support education reform and lead to better teaching and learning in schools. In pursuit of this aim, it conducts, and reports on, major studies of student achievement in literacy, mathematics, science, citizenship, and digital literacy. These studies, most notably TIMSS, PIRLS, ICCS, and ICILS, are well established and have set the benchmark for international comparative studies in education.

The studies have generated vast datasets encompassing student achievement, disaggregated in a variety of ways, along with a wealth of contextual information which contains considerable explanatory power. The numerous reports that have emerged from them are a valuable contribution to the corpus of educational research.

Valuable though these detailed reports are, IEA's goal of supporting education reform needs something more: deep understanding of education systems and the many factors that bear on student learning advances through in-depth analysis of the global datasets. IEA has long championed such analysis and facilitates scholars and policymakers in conducting a secondary analysis of our datasets. So, we provide software such as the International Database Analyzer to encourage the analysis of our datasets, support numerous publications including a peer-reviewed journal—*Large-scale Assessment in Education*—dedicated to the science of large-scale assessment and publishing articles that draw on large-scale assessment databases, and organize a biennial international research conference to nurture exchanges between researchers working with IEA data.

The **IEA Research for Education** series represents a further effort by IEA to capitalize on our unique datasets, so as to provide powerful information for policymakers and researchers. Each report focuses on a specific topic and is produced by a dedicated team of leading scholars on the theme in question. Teams are selected on the basis of an open call for tenders; there are two such calls a year. Tenders are subject to a thorough review process, as are the reports produced (Full details are available on the IEA website).

This ninth volume in the series addresses student misconceptions and errors in physics and mathematics. Student error is fertile ground for research and can yield rich material for pedagogical improvement. IEA has long espoused the benefits of error analysis, and many countries have conducted error analyses of national datasets within the Trends in International Mathematics and Science Study (TIMSS).

This book reports on a study that examined student misconceptions and errors across education systems and over time. Specifically, it draws on 20 years of data (1995–2015) from TIMSS at grades four and eight, and from TIMSS Advanced (grade 12), looking at five countries (Italy, Norway, Russian Federation, Slovenia, and the USA) that participated in TIMSS across the 20-year period. To permit in-depth focus, the study is restricted to one topic in physics (gravity) and one in mathematics (linear equations); these topics were chosen because they were covered in increasing depth across the grades, and assessment items were available for them within each of the three assessments.

The authors have used relatively straightforward analyses to understand the way in which students engage with the test and the way that error patterns manifest. In common with similar studies, the work is based on item-level performance data for individual test items but goes beyond this by focusing on sets of items that span grade level and education system. The data was also analyzed by gender. The authors provide detail on the nature and distribution of student errors and how they vary by grade level, gender, and country over the 20-year period.

This book will be a valuable resource for teachers and teacher educators on how best to teach these topics so as to enhance student learning. Moreover, the methodology deployed here can be used to investigate student misconceptions and errors in a variety of other topics.

Seamus Hegarty
Leslie Rutkowski
Series Editors

Contents

Chapter 1
An Introduction to Student Misconceptions and Errors in Physics and Mathematics

Abstract For the past few decades, the focus on science, technology, engineering, and mathematics (STEM) education has grown, with many countries seeking to increase the number of students who pursue further study and careers in STEM. It is thus important to identify which science and mathematics concepts are problematic for students to determine what changes may be needed in the school curricula to improve the teaching and learning of these key subjects throughout elementary, middle, and secondary school. The research in this book investigates patterns of student misconceptions, errors, and misunderstandings across education systems, grade levels, gender, and time using 20 years of data from the Trends in International Mathematics and Science Study (TIMSS) and TIMSS Advanced assessments (1995–2015). Students' level of understanding of selected physics and mathematics topics (gravity and linear equations) were assessed using data from the TIMSS assessments at grades four and eight, and the TIMSS Advanced assessments of students in their final year of secondary school taking advanced coursework in physics and mathematics. Diagnostic item-level performance data were used to trace student misconceptions, errors, and misunderstandings related to these topics across grade levels. The patterns in misconceptions may inform instruction by identifying specific gaps or deficiencies in the curricula across grade levels.

Keywords Assessment framework objective · Diagnostic data · Errors · Gender differences · Gravity · International large-scale assessment · Item-level data · Linear equations · Mathematics · Misconceptions · Physics · Science · Student achievement · Trend analysis · Trends in International Mathematics and Science Study (TIMSS) · Italy · Norway · Russian Federation · Slovenia · United States

1.1 Introduction

With the increasing emphasis on science, technology, engineering, and mathematics (STEM) education and careers, it is important to assess students throughout their education in the core subjects of mathematics and science, and identify

T. Neidorf et al., *Student Misconceptions and Errors in Physics and Mathematics*, IEA Research for Education 9, https://doi.org/10.1007/978-3-030-30188-0_1

persisting student misconceptions, errors, and misunderstandings. Understanding how misconceptions, errors, and misunderstandings in the higher grade levels relate to a lack of foundational understanding at earlier grades is important for many stakeholders in science and mathematics education, including classroom teachers, teacher educators, policymakers, and researchers. This report analyzes specific student misconceptions, errors, and misunderstandings related to core physics and mathematics concepts; the results may inform improvements in the teaching, learning, and reinforcement of these core concepts throughout elementary, middle, and secondary school.

We used assessment items and student performance data from the Trends in International Mathematics and Science Study (TIMSS) and TIMSS Advanced assessments conducted across 20 years (1995–2015)[1] to explore students' level of understanding of two core topics (gravity and linear equations), and the nature and extent of their misconceptions, errors, and misunderstandings at grade four and grade eight (TIMSS students), and in the final year of secondary school (TIMSS Advanced). We report results for five countries that participated in the TIMSS Advanced 2015 assessment, namely Italy, Norway, the Russian Federation, Slovenia, and the United States. These countries were selected from the nine countries that participated in TIMSS Advanced 2015 as they also participated in all, or nearly all, TIMSS grade four and grade eight assessments from 1995 to 2015. The data thus maximize the number of comparisons across countries and grade levels, and enable us to report performance patterns over time across multiple assessment cycles.[2] The other four countries that participated in TIMSS Advanced 2015 (France, Lebanon, Portugal, and Sweden) did not participate in TIMSS 2015 at both grades four and eight, or had missing data for more than one prior assessment cycle for at least one grade level. The specific assessments in which each country participated are summarized in Chap. 3.

Using TIMSS and TIMSS Advanced assessment data to explore student misconceptions, errors, and misunderstandings has multiple advantages. First, the TIMSS and TIMSS Advanced assessments have been administered to nationally representative samples of students at regular intervals, starting in 1995 (with the most recent assessments conducted in 2015).[3] In contrast, most research studies

[1]The Trends in International Mathematics and Science Study (TIMSS) is a flagship study of the International Association for the Evaluation of Educational Achievement (IEA), coordinated by the world-renowned TIMSS & PIRLS International Study Centre at Boston College. TIMSS and TIMSS Advanced are international comparative studies designed to measure trends in mathematics and science achievement and collect information about educational contexts that may be related to student achievement. As in all IEA studies, the international coordination is carried out in cooperation with the national research coordinators in each participating education system. For more information about TIMSS and TIMSS Advanced, see www.iea.nl/timss.

[2]Although our study focuses on these specific countries, the methodology described can be applied to an individual education system or any set of education systems.

[3]TIMSS has been administered every four years, starting in 1995 (although the 1999 assessment was administered at grade eight only), and TIMSS Advanced was administered in 1995, 2008, and 2015.

investigating student misconceptions use fairly small samples from a particular region, district, or school (Alonzo et al. 2012) and are conducted within a limited time frame. Second, TIMSS provides the ability to track performance of student cohorts at three grade levels across multiple assessment years, permitting the evaluation of student performance and misconceptions over time. Lastly, TIMSS and TIMSS Advanced provide access to sets of released items (questions from the assessments) and student performance data from each assessment cycle that can be used for research purposes, such as the diagnostic item-level results in this report. The results may provide a more comprehensive picture of student performance within and across countries.

TIMSS and TIMSS Advanced data have been used in a number of secondary analyses conducted to address the topic of student misconceptions in different countries (Angell 2004; Juan et al. 2017; Mosimege et al. 2017; Prinsloo et al. 2017; Provasnik et al. 2019; Saputro et al. 2018; Văcăreţu n.d.; Yung 2006). Following the release of the 2015 TIMSS and TIMSS Advanced results in the United States (Provasnik et al. 2016), the American Institutes for Research (AIR) conducted in-depth secondary analyses of TIMSS and TIMSS Advanced data from the United States. An initial report on the United States' performance in TIMSS Advanced 2015 described areas of relative strength and weakness, and common approaches, misconceptions, and errors in advanced mathematics and physics (Provasnik et al. 2019). A follow-up study using both TIMSS and TIMSS Advanced data further explored how physics misconceptions demonstrated by TIMSS Advanced students in the United States can be traced back to misconceptions, or a lack of foundational understanding about physics concepts in earlier grades (unpublished work).[4]

In this report, we expand upon such previous work and describe the methodology we use to (1) investigate misconceptions, errors, and misunderstandings in both physics and mathematics; (2) explore patterns of misconceptions, errors, and misunderstandings across grade levels for a select group of countries; (3) report differences in these patterns across countries, overall and by gender; and (4) report differences across assessment years.

1.2 Defining the Terminology

To begin, we first define the terms used throughout this report as they apply to physics and mathematics.[5]

[4]Presented at the 2018 annual conference of the National Association for Research in Science Teaching (NARST), Atlanta, GA.

[5]See Sect. 3.2.3 for further information about the methods and rationales for the treatment of different response types (incorrect, off-task, and blank) under misconceptions, errors, and misunderstandings.

1.2.1 Performance Objectives

Performance objectives are based on the set of TIMSS and TIMSS Advanced items selected for the study. They describe the specific knowledge and abilities expected of students at different grade levels (i.e., what they must know and be able to do in order to respond correctly to the TIMSS and TIMSS Advanced assessment items). For this report, there are four performance objectives identified related to gravity and nine related to linear equations, each measured by one or more assessment items (see Chap. 4). Some performance objectives were assessed at only one grade level, while others were assessed by items in two grade levels (i.e., TIMSS Advanced/grade eight or grade eight/grade four) or in all three grade levels (for physics only).

1.2.2 Misconceptions in Physics

Misconceptions apply only to the physics items. These reflect students' incorrect preconceived notions about a physics concept, usually based on their experiences or observations of physical phenomena in daily life. In this report, a misconception is demonstrated by particular types of student responses such as specific incorrect response options for multiple-choice items or specific incorrect scoring guide categories for constructed-response items (where students provide a written response).

1.2.3 Errors in Mathematics

Errors apply only to mathematics items where students are expected to follow a set mathematical procedure to obtain the correct response. Errors reflect any type of response where the correct answer was not obtained.

1.2.4 Misunderstandings in Physics and Mathematics

Misunderstandings can apply to both physics and mathematics items. These reflect responses where students did not demonstrate that they understood the physics or mathematics concept as it applies to the item, but do not involve procedural errors in mathematics or signify a specific misconception in physics as defined above.

Physics Items
Includes items (mostly constructed-response) where students must apply their understanding of the physics concept to a given situation, but specific incorrect response types are not tracked. Misunderstandings in physics indicate a lack of

understanding and include all incorrect responses (including off-task and blank responses).

Mathematics Items
Includes items where there is no set procedure required, and students must figure out how to apply their understanding of the mathematics concept to answer the question. A misunderstanding in mathematics may be demonstrated by specific types of incorrect student responses or by all incorrect responses (including off-task and blank responses).

1.3 Core Concepts in Physics and Mathematics

We focus on core concepts in physics and mathematics that are introduced in elementary school and further developed across grades through middle school and secondary school. To fully demonstrate our methodology for exploring students' misconceptions, errors, and misunderstandings across grade levels, we selected the specific topics of gravity in physics and linear equations in mathematics. These topics reflect key concepts that are covered in both the TIMSS and TIMSS Advanced assessment frameworks, and there are items covering these topics (or their precursors) in the grade four and eight assessments, and the TIMSS Advanced assessment. This allowed us to trace misconceptions, errors, and misunderstandings across all three grade levels.

Gravity is a fundamental concept introduced to students at an early age, and students enter school with preconceptions about the topic based on their experiences and observations of physical phenomena in their daily life. The topic is covered in physical science, earth science, and more advanced physics courses in secondary school, and the depth of understanding related to gravity is expected to develop across the grades. The topic of gravity (gravitational force) provides a good context for evaluating students' abilities to apply force concepts, and can be used to identify some general misconceptions related to force and motion across all three grade levels.

Based on the TIMSS 2015 frameworks (Jones et al. 2013), students at grade four can identify gravity as the force that draws objects toward Earth and recognize that forces may cause an object to change its motion (Table 1.1). At grade eight, students can describe common mechanical forces, including gravitational force, acting on objects and can predict the changes in motion (if any) of an object based on the forces acting on it. In addition, by grade eight, students recognize that it is the force of gravity that keeps planets and moons in orbit and pulls objects to Earth's surface. The 2015 TIMSS Advanced physics framework (Jones et al. 2014) expects students at the end of secondary school to use Newton's laws of motion to explain the dynamics of different types of motion and how the action of combined forces influences a body's motion.

Table 1.1 TIMSS 2015 and TIMSS Advanced 2015 assessment framework objectives related to gravity

TIMSS grade 4	TIMSS grade 8	TIMSS Advanced physics
Physical science • Identify gravity as the force that draws objects to Earth • Recognize that forces (pushing and pulling) may cause an object to change its motion and compare the effects of forces of different strengths in the same or opposite direction acting on an object	*Physics* • Describe common mechanical forces, including gravitational, normal, friction, elastic, and buoyant forces, and weight as a force due to gravity • Predict qualitative one-dimensional changes in motion (if any) of an object based on the forces acting on it *Earth science* • Recognize that it is the force of gravity that keeps the planets and moons in orbits as well as pulls objects to Earth's surface	*Mechanics* • Predict and determine the position, displacement, and velocity of bodies given initial conditions; and use Newton's laws of motion to explain the dynamics of different types of motion and to calculate displacement, velocity, acceleration, distance traveled, or time elapsed • Identify forces, including frictional force, acting on a body at rest, moving with constant velocity, or moving with constant acceleration and explain how their combined action influences the body's motion; and find solutions to problems involving forces

Notes This outlines the portion of the objectives included in the 2015 TIMSS and TIMSS Advanced frameworks that specifically relate to the physics concepts and assessment items discussed in this report
Source International Association for the Evaluation of Educational Achievement (IEA), Trends in International Mathematics and Science Study (TIMSS) 2015 and TIMSS Advanced 2015 assessment frameworks (Jones et al. 2013, 2014)

For mathematics, we focused on linear equations for several reasons. Algebra, and the topic of linear equations specifically, spans students' mathematics education in elementary, middle school, and secondary school. In the 2015 TIMSS mathematics framework (Grønmo et al. 2013), students at grade four can identify or write expressions or number sentences to represent problem situations involving unknowns; identify and use relationships in well-defined patterns; solve problems set in contexts; and read, compare, and represent data from tables and line graphs (Table 1.2). At grade eight, students can write equations or inequalities to represent situations; solve simultaneous linear equations in two variables; interpret, relate, and generate representations of functions in tables, graphs, or words; and interpret the meanings of slope and y-intercept in linear functions. The 2015 TIMSS Advanced mathematics framework (Grønmo et al. 2014) expects students at the end of secondary school to solve linear and quadratic equations, as well as systems of linear equations and inequalities, and to use equations and inequalities to solve contextual problems.

Not only do students continue to study the topic of linear equations across grades, their conceptual understanding of linear equations progresses from concrete (number

Table 1.2 TIMSS 2015 and TIMSS Advanced 2015 assessment framework objectives related to linear equations: 2015

TIMSS grade 4	TIMSS grade 8	TIMSS Advanced mathematics
Number • Identify or write expressions or number sentences to represent problem situations involving unknowns • Identify and use relationships in a well-defined pattern (e.g., describe the relationship between adjacent terms and generate pairs of whole numbers given a rule) • Solve problems set in contexts, including those involving measurements, money, and simple proportions *Data* • Read, compare, and represent data from tables, and line graphs	*Algebra* • Write equations or inequalities to represent situations • Solve linear equations, linear inequalities, and simultaneous linear equations in two variables • Interpret, relate, and generate representations of functions in tables, graphs, or words • Interpret the meanings of slope and y-intercept in linear functions	*Algebra* • Solve linear and quadratic equations and inequalities as well as systems of linear equations and inequalities • Use equations and inequalities to solve contextual problems

Notes This outlines the portion of the objectives included in the 2015 TIMSS and TIMSS Advanced frameworks that specifically relate to the mathematics concepts and assessment items discussed in this report
Source International Association for the Evaluation of Educational Achievement (IEA), Trends in International Mathematics and Science Study (TIMSS) 2015 and TIMSS Advanced 2015 assessment frameworks (Grønmo et al. 2013, 2014)

sentences at grade four) to abstract (equations and graphical representations at grade eight and the upper secondary level) as their mathematics competency progresses. In addition, students' performance in algebra is linked to higher achievement in mathematics (Walston and McCarroll 2010). The topic of linear equations is one of the most basic in algebra because linear equations are much simpler than other types of relationships, such as quadratic and exponential equations. Before students can understand characteristics like intercepts and slope in these more complex relationships, they must master the same characteristics in linear equations. Finally, the topic of linear equations is versatile in terms of connecting mathematics to other subject areas and real-world applications. For example, understanding graphs of equations is an integral skill in science classrooms. Similarly, understanding equations is important in general life skills, including all aspects of financial literacy. The focus on linear equations will, therefore, provide an examination of students' performance in a topic area that is important for postsecondary success.

1.4 Research Questions

Our methodology (see Chap. 3) includes three major components: (1) assessment framework review and content mapping to identify the set of items measuring the topics of interest at each grade level; (2) evaluation of diagnostic item-level performance data to identify the specific performance objectives measured by these items and to provide evidence of specific misconceptions, errors, and misunderstandings; and (3) analyses of the percentage of students demonstrating these misconceptions, errors, and misunderstandings across countries by grade level, gender, and assessment year. Example items are shown in the report to illustrate the specific types of misconceptions, errors, and misunderstandings demonstrated by students at each grade level.[6]

Using item-level performance data from multiple assessment cycles of TIMSS and TIMSS Advanced (from 1995 to 2015), we addressed three research questions.

Research question 1: What are common types of student misconceptions, errors, and misunderstandings in grade four, grade eight, and the final year of secondary school (TIMSS Advanced students), and how do they compare across countries?

For each selected country, we determined the frequency of specific types of misconceptions, errors, and misunderstandings related to gravity and linear equations demonstrated on items from across the three grade levels, and identified and compared patterns across countries and grade levels.

Research question 2: How do student misconceptions, errors, and misunderstandings differ by gender?

For each assessment item, we determined differences in student performance and the frequency of specific types of misconceptions, errors, and misunderstandings by gender, and compared these differences across countries and grade levels.

Research question 3: How persistent are patterns in misconceptions, errors, and misunderstandings over time?

Using trend items administered in multiple assessment cycles, we compared the frequency of specific types of misconceptions, errors, and misunderstandings across all of the TIMSS assessments conducted between 1995 and 2015 to discover whether patterns across countries changed over time (e.g., did specific

[6]Example items shown in this report include "restricted-use" items from the TIMSS 2015 assessments and released items from prior assessment years. The 2015 "restricted-use" items are those designated for use as examples in the international reports and by participating countries in their national reports or for secondary research. Although example items are limited to released or restricted-use items, appropriate non-released (secure) items from 2015 were included in the analyses of misconceptions but are not shown in the report. All example items ("restricted-use" and "released") are shown with permission from IEA.

misconceptions at grades four or eight increase, decrease, or stay the same between 2007 and 2015?).[7]

While this report is focused on student misconceptions, errors, and misunderstandings related to the two topics of gravity and linear equations, the general methodology described in the report can be applied to a range of mathematics and science topics covered in TIMSS and TIMSS Advanced. This methodology can be used to trace misconceptions across all three grade levels (as in this report) or two grade levels (e.g., grade eight and grade four), or it can be used to focus on patterns of misconceptions at one grade only. The results can inform instruction across grades, by relating country-level patterns in misconceptions, errors, and misunderstandings to specific gaps or deficiencies in the curricula.

References

Alonzo, A. C., Neidorf, T., & Anderson, C. W. (2012). Using learning progressions to inform large-scale assessment. In A. C. Alonzo, & A. W. Gotwals (Eds.), *Learning progressions in science*. Rotterdam, The Netherlands: Sense Publishers.

Angell, C. (2004). *Exploring students' intuitive ideas based on physics items in TIMSS-1995*. In C. Papanastasiou (Ed.), *Proceedings of the IRC-2004 TIMSS. IEA International Research Conference* (Vol. 2, pp. 108–123). Nicosia, Cyprus: University of Cyprus. Retrieved from https://www.iea.nl/sites/default/files/2019-03/IRC2004_Angell.pdf.

Grønmo, L. S., Lindquist, M., & Arora, A. (2014). TIMSS Advanced 2015 mathematics framework. In I.V. S. Mullis, & M. O. Martin (Eds.), *TIMSS Advanced 2015 assessment frameworks* (pp. 9–15). Chestnut Hill, MA: TIMSS & PIRLS International Study Center, Boston College. Retrieved from https://timss.bc.edu/timss2015-advanced/downloads/TA15_FW_Chap1.pdf.

Grønmo, L. S., Lindquist, M., Arora, A., & Mullis, I.V. S. (2013). TIMSS 2015 mathematics framework. In I.V. S. Mullis, & M. O. Martin (Eds.), *TIMSS 2015 assessment frameworks* (pp. 11–27). Chestnut Hill, MA: TIMSS & PIRLS International Study Center, Boston College. Retrieved from https://timssandpirls.bc.edu/timss2015/downloads/T15_FW_Chap1.pdf.

Jones, L. R., Wheeler, G., & Centurino, V. A. S. (2013). TIMSS 2015 science framework. In I.V. S. Mullis, & M. O. Martin (Eds.) *TIMSS 2015 assessment frameworks* (pp. 29–58). Chestnut Hill, MA: TIMSS & PIRLS International Study Center, Boston College. Retrieved from https://timssandpirls.bc.edu/timss2015/downloads/T15_FW_Chap2.pdf.

Jones, L. R., Wheeler, G., & Centurino, V. A. S (2014). TIMSS Advanced 2015 physics framework. In I. V. S. Mullis and M. O. Martin (Eds.), *TIMSS Advanced 2015 assessment frameworks* (pp. 17–25). Chestnut Hill, MA: TIMSS & PIRLS International Study Center, Boston College. Retrieved from https://timss.bc.edu/timss2015-advanced/downloads/TA15_FW_Chap2.pdf.

Juan, A., Hannan, S., Zulu, N., Harvey, J. C., Prinsloo, C. H., Mosimege, M., & Beku, U. (2017). *TIMSS item diagnostic report: South Africa: Grade 5 Numeracy. (Commissioned by the Department of Basic Education)*. South Africa: Human Sciences Research Council. Retrieved from http://ecommons.hsrc.ac.za/handle/20.500.11910/11447.

[7]Trends in misconceptions, errors, and misunderstandings are only reported for grade four and grade eight, as there were no trend items measuring the selected topics in TIMSS Advanced.

Mosimege, M., Beku, U., Juan, A., Hannan, S., Prinsloo, C.H., Harvey, J.C., & Zulu, N. (2017). *TIMSS item diagnostic report: South Africa: Grade 9 Mathematics. (Commissioned by the Department of Basic Education)*. South Africa: Human Sciences Research Council. Retrieved from http://repository.hsrc.ac.za/handle/20.500.11910/11448.

Prinsloo, C. H., Harvey, J. C., Mosimege, M., Beku, U., Juan, A., Hannan, S., & Zulu, N. (2017). *TIMSS item diagnostic report: South Africa: Grade 9 Science. (Commissioned by the Department of Basic Education)*. South Africa: Human Sciences Research Council. Retrieved from http://www.hsrc.ac.za/en/research-data/ktree-doc/19286.

Provasnik, S., Malley, L., Neidorf, T, Arora, A., Stephens, M., Balestreri, K., Perkins, R., & Tang, J. H. (2019, in press). *U.S. performance on the 2015 TIMSS Advanced mathematics and physics assessments: A closer look (NCES 2017-020)*. Washington, DC: US Department of Education, National Center for Education Statistics.

Provasnik, S., Malley, L., Stephens, M., Landeros, K., Perkins, R., & Tang, J. H. (2016). *Highlights from TIMSS and TIMSS Advanced 2015: Mathematics and science achievement of U.S. students in grades 4 and 8 and in advanced courses at the end of high school* (NCES 2017-002). Washington, DC: US Department of Education, National Center for Education Statistics. Retrieved from https://nces.ed.gov/pubs2017/2017002.pdf.

Saputro, B. A., Suryadi, D., Rosjanuardi, R., & Kartasasmita, B. G. (2018). Analysis of students' errors in responding to TIMSS domain algebra problem. *Journal of Physics: Conference Series, 1088*, 012031. Retrieved from https://iopscience.iop.org/article/10.1088/1742-6596/1088/1/012031.

Văcăreţu, A. (n.d.). *Using the TIMSS results for improving mathematics learning*. Cluj-Napoca, Romania: Romanian Reading and Writing for Critical Thinking Association. Retrieved from http://directorymathsed.net/montenegro/Vacaretu.pdf.

Walston, J., & McCarroll, J. C. (2010). Eighth-grade algebra: Findings from the eighth-grade round of the early childhood longitudinal study, kindergarten class of 1998–99 (ECLS-K). *Statistics in Brief, 16*, 1–20.

Yung, B. H. W. (Ed.). (2006). *Learning from TIMSS: Implications for teaching and learning science at the junior secondary level*. TIMSS HK IEA Centre. Hong Kong: Education and Manpower Bureau. Retrieved from https://cd1.edb.hkedcity.net/cd/science/is/timss/learningfromtimss.pdf.

Chapter 2
Review of Research into Misconceptions and Misunderstandings in Physics and Mathematics

Abstract Many diagnostic methods have been used to analyze data from large-scale assessments such as the Trends in International Mathematics and Science Study (TIMSS), and the prior research on misconceptions and misunderstandings in physics and mathematics is extensive. This literature review provides an overview of different diagnostic models that have been used to explore student attributes and misconceptions in mathematics and science and how they compare to the methodology used in this study. A comprehensive review of prior research into student misconceptions and misunderstandings in physics related to gravitational force and in mathematics related to linear equations connects the established literature to the current study.

Keywords Diagnostic models · Errors · Gravity · International large-scale assessment · Linear equations · Mathematics · Misconceptions · Physics · Science · Student achievement · Trend analysis · Trends in International Mathematics and Science Study (TIMSS)

2.1 Introduction

When measuring student achievement, traditional methods of analysis often focus on what students know (i.e., the correct answers). For example, large-scale assessments such as IEA's TIMSS use unidimensional models such as item response theory (IRT) to measure individual students' latent abilities, skills, and knowledge. Recent research using multidimensional models has begun to consider both correct and incorrect patterns when measuring and reporting on specific skills/abilities and misconceptions. Prior research has highlighted the importance of identifying and understanding student misconceptions to improve learning in both physics and mathematics.

We divide the literature review into three sections. The first section reviews the variety of diagnostic models that have been used to explore student attributes and misconceptions, misunderstandings, and errors in mathematics and science. The

T. Neidorf et al., *Student Misconceptions and Errors in Physics and Mathematics*, IEA Research for Education 9, https://doi.org/10.1007/978-3-030-30188-0_2

second and third sections explore prior research into student misconceptions, misunderstandings, and errors in physics related to gravitational force, and in mathematics related to linear equations, respectively. Both sections also look at gender differences in the prevalence of misconceptions.

2.2 Diagnostic Models Overview

Traditional psychometric models used for test analysis, such as IRT models, often focus on measuring a single latent continuum representing overall ability (Bradshaw and Templin 2014). Although these models are considered an important means of assessing student knowledge, their focus on measuring one underlying student ability is limiting. De la Torre and Minchen (2014) noted that the unidimensional nature of these methods made them less effective as diagnostic models. The need for models that would provide diagnostic information spurred the development of a new class of test models known as cognitive diagnostic models (CDMs).

A CDM is a type of model that classifies different combinations of mastered student attributes into different latent classes. It then determines students' abilities based on various skills or attributes that students have or have not mastered (de la Torre and Minchen 2014; Henson et al. 2009). An example of a CDM model is the diagnostic classification model (DCM), which uses distractor-driven tests (designed to measure both "desirable and problematic aspects of student reasoning") or multiple-choice tests that measure multidimensional attributes (Shear and Roussos 2017). In addition to the DCM, there are many other types of CDMs, such as the rule space model (Tatsuoka 1983), the deterministic input, noisy "and" gate (DINA) model (Junker and Sijtsma 2001), the noisy input, deterministic "and" gate (NIDA) model (Maris 1999), and the reparametrized unified model (RUM) (Roussos et al. 2007). Each of these models vary in terms of their complexity, the parameters they assign to each item, and the assumptions made when random noise enters the test-taking process (Huebner and Wang 2011). The varied and multidimensional nature of CDMs makes them better suited to performing educational diagnoses. In fact, a recent study by Yamaguchi and Okada (2018) using TIMSS 2007 mathematics data found that CDMs had a better fit than IRT models.

A relatively new approach, the scaling individuals and classifying misconceptions (SICM) model, investigated by Bradshaw and Templin (2014), combines the IRT model and the DCM to provide a statistical tool to measure misconceptions. The SICM model uses data on wrong answers by modeling categorical latent variables that represent "misconceptions" instead of skills. To categorize misconceptions, the authors cited inventories such as the force concept inventory (Hestenes et al. 1992), an assessment of the Newtonian concept of force.

For large-scale assessments, such as TIMSS, applying these current diagnostic models can be difficult since the TIMSS assessments were not designed as cognitive diagnostic assessments that measure specific components of skills/abilities, nor were they designed using a CDM with pre-defined attributes (de la Torre and Minchen 2014; Leighton and Gierl 2007). However, some studies have shown that applying these approaches to TIMSS data can provide valuable information about test takers. Dogan and Tatsuoka (2008) used the rule space model to evaluate Turkish performance on the TIMSS 1999 grade eight mathematics assessment (also known as the Third International Mathematics and Science Study-Repeat, or TIMSS-R), determining that Turkish students demonstrated weaknesses in skills such as applying rules in algebra and quantitative reading. Another study (Choi et al. 2015) also used a CDM approach to compare performance on the TIMSS mathematics assessment between the United States and Korean grade eight samples. While these studies showed that CDM can offer valuable information on student concept mastery in TIMSS, these studies also acknowledged there are limitations when applying these models to this assessment.

In general, CDMs and SICMs use best-fit models to predict student-level proficiency and misconceptions, and these models would be most efficient when used on computer adaptive tests (CATs), so that "all test takers can be measured with the same degree of precision" (Hsu et al. 2013). The TIMSS assessments, which are not designed for student-level reporting and are not computer-adaptive, are not catered to CDMs and SICMs. Based on the TIMSS assessment design, only a portion of the items are administered to each student; thus, the claims that can be made about student proficiency on specific skills and concepts are limited.[1]

In contrast to research using the types of diagnostic models described above, our study used a different diagnostic approach based on item-level performance data (i.e., frequency distributions across response categories) for individual assessment items to explore the nature and extent of students' misconceptions, errors, and misunderstandings demonstrated by their incorrect responses. Other studies conducted by countries participating in TIMSS have taken a similar approach to describing student understanding and misconceptions based on their responses to individual TIMSS and TIMSS Advanced mathematics and science items at different grade levels (Angell 2004; Juan et al. 2017; Mosimege et al. 2017; Prinsloo et al. 2017; Provasnik et al. 2019; Saputro et al. 2018; Văcăreţu, n.d.; Yung 2006). For example, Angell (2004) analyzed student performance on TIMSS Advanced 1995 physics items in Norway; a series of diagnostic reports published in South Africa used item-level data from TIMSS 2015 to describe performance of their students in mathematics for grade five (Juan et al. 2017) and grade nine (Mosimege et al. 2017), and in science for grade nine (Prinsloo et al. 2017); and Saputro et al. (2018) used performance on algebra items from TIMSS 2011 to understand the types of errors made by students in Indonesia. All of these reports presented released items from TIMSS and TIMSS Advanced and described common types of incorrect answers given by students on the assessments, finding that misconceptions were often context-dependent and could be missed in broader analyses.

---

[1]TIMSS uses a matrix-sampling design whereby a student is administered only a sample of the assessment items; most items are missing by design for each student.

Our study goes beyond looking at individual assessment items by focusing on sets of items that measure specific concepts of interest in physics and mathematics across grade levels (gravity and linear equations, in this case). Student performance on these items are used to report on patterns in misconceptions across countries, grades, and assessment cycles, and by gender. Considering the assessment design of TIMSS, there is unique value in this approach to focus on item-level data to make country-level inferences and better understand how student misconceptions have changed over time in different cultural contexts.

2.3 Misconceptions in Physics

Physics misconceptions (including those related to gravity) held by students of varying ages have been studied extensively. Previous research has included investigations of primary, secondary, and university students (Darling 2012; Demirci 2005; Hestenes et al. 1992; Pablico 2010; Piburn et al. 1988; Stein et al. 2008), as well as pre-service teachers (Gönen 2008). The literature about misconceptions related to gravitational force demonstrates that alternate conceptions of physical observations and processes based on intuition or preconceived notions are common and pervasive.

When analyzing misconceptions in physics, many researchers have focused on "common sense beliefs," a "system of beliefs and intuitions about physical phenomena derived from extensive personal experience" that students may develop before they even enter the classroom (Halloun and Hestenes 1985a, b). Many of these beliefs are misconceptions inconsistent with scientific explanations provided during formal instruction; moreover, they are difficult to overcome and can inhibit students from understanding and applying more advanced physics concepts if not addressed early on. Numerous studies have been conducted to further explain these misunderstandings and several diagnostic tests have been developed to measure them, the most widely used being the force concept inventory, which uses multiple-choice items to track student misconceptions relating to "common sense beliefs" (Hestenes et al. 1992). Research has shown that many physics misconceptions are best overcome by focused instruction that actively aims to address these misconceptions (Eryilmaz 2002; Hestenes et al. 1992; Thornton et al. 2009).

Misconceptions based on common-sense beliefs tend to be incompatible with many physics concepts, such as Newton's laws. For example, several studies have documented that students believe that there is always a force in the direction of motion and that this belief sometimes prevails even after college instruction (Clement 1982; Hestenes et al. 1992; Thornton and Sokoloff 1998). Another well-documented misconception is that it is not possible to have acceleration without velocity (Kim and Pak 2002; Reif and Allen 1992). These misconceptions can often stem from students' inability to distinguish between velocity, acceleration, and force (Reif and Allen 1992; Trowbridge and McDermott 1980). In particular, many students struggle with gravitational force. The concept appears to be poorly learned at the secondary level, with related misconceptions continuing in higher levels of education (Bar et al. 2016; Kavanaugh and Sneider 2007).

In addition, many students' conceptions of gravity are closely related to their conceptions of a spherical Earth (Gönen 2008; Nussbaum 1979; Sneider and Pulos 1983). When conducting interviews with children in grades six and 10 on what objects presented to them were acted on by gravity, Palmer (2001) found that <30% of students in each grade level were able to correctly answer that all of the objects were acted on by gravity. Some students, Palmer noted, also believed that buried objects (beneath the surface of Earth) were not subject to gravity.

Many of these misconceptions have been shown to be stable in the face of conventional physics instruction, preventing students from learning new concepts. One previous study on misconceptions about force and gravity investigated high school students' conceptions about the direction of motion and force on a ball being thrown upward and then falling back down (Pablico 2010). The majority of students in the study (grades 9–12) demonstrated the misconception that the net force on the ball was always in the direction of motion throughout the ball's path, not understanding that it is the constant downward force due to gravity that causes the observed changes in motion. Many students thought that the force was directed upward during the ball's upward motion and that the force was zero when the ball was at the top of its flight (when it stops momentarily and changes direction). Although students identified the force as downward when the ball was traveling down, most were not able to correctly justify this answer, with many students believing that the force must be directed down since the ball is moving downward.

Other research has described instances of gender gaps in students' understanding in physics. For example, at the beginning of physics courses, females tend to start with lower levels of conceptual understanding, and conventional instructional approaches are not effective in shrinking this gender gap (Cavallo et al. 2004; Docktor and Heller 2008; Hake 2002; Hazari et al. 2007; Kost et al. 2009).

2.4 Misunderstandings in Mathematics

In mathematics, algebra is often considered a gatekeeper to higher education and related career paths (Kilpatrick and Izsák 2008). Although algebraic understanding is considered crucial for student success in more advanced mathematics courses, many scholars have documented that students struggle with algebraic concepts, especially those relating to linear equations.

Solving linear equations requires a balance of conceptual knowledge and procedural skills. Conceptual knowledge involves having an understanding of principles and relationships, while procedural skills involve the ability to carry out a sequence of operations effectively (Gilmore et al. 2017). Unlike simpler arithmetic problems, solving linear equations involves much more than merely memorizing and applying a formula to solve an equation; it also includes understanding the relationship between the quantities represented. Conceptually, students need a deep understanding of independent and dependent variables to explain what slope or intercepts mean in a given situation (Kalchman and Koedinger 2005). Yet many students have shown a tendency to rely on procedural knowledge despite lacking a conceptual understanding of the equation (Caglayan and Olive 2010).

Stump (2001) argued that although high school pre-calculus students have been exposed to formal instruction, their conceptual understanding of "slope" is not well developed. When testing a group of high schoolers in her study, Stump found that many students understood slope in functional situations but were unable to recognize it as a measure of rate of change or as a measure of steepness. Other researchers noted that while gaining an understanding of slope, students they interviewed were unable to recognize the difference between additive and multiplicative relationships (Simon and Blume 1994) or were unable to understand ratio as a measure of slope (Swafford and Langrall 2000). This inability to develop a conceptual knowledge of the relationship between variables has contributed to many misunderstandings related to slope and linear equations.

Lack of conceptual knowledge about the relationship between variables in linear equations also impacts a student's ability to understand and translate the symbolic nature of linear equations. Official standards, such as those of the National Council of Teachers of Mathematics (NCTM), recommend that students must be able to "represent and analyze relationships using tables, verbal rules, equations, and graphs" (NCTM 1989). Yet many students find it very difficult to represent equations graphically. Research suggests that this is because students tend to lack a strong understanding of the relationship between algebraic equations and graphical representations (Knuth 2000).

Even when using a graphical approach would ensure a higher likelihood of success, researchers have found that students were reluctant to use graphs (Knuth 2000; Tsamir and Almog 2001; Dyke and White 2004). For example, Knuth (2000) found that even when working on problems designed to encourage the use of graphical reasoning, students demonstrated a strong reliance on other solution methods and failed to use graphical-solution methods. In another study, Huntley et al. (2007) conducted clinical interviews of third year high school mathematic students and found that many students needed to be prompted to use graphical solutions even it was the most efficient method to solve the equation.

This difficulty with modeling algebraic relationships graphically makes it difficult for students to translate real life word problems into the appropriate algebraic equations (Adu et al. 2015; Bishop et al. 2008). Without focused and deliberate instruction, it would be difficult for students to tackle these algebraic misunderstandings as they progress to higher levels of mathematics. As noted in the physics section, some research in this area has found that males make fewer mistakes than females and make different types of mistakes when solving problems related to multi-step linear equations in algebra (Powell 2013).

This report contributes to the literature on research into students' misconceptions and misunderstandings in physics and mathematics by studying specific types of related misconceptions, errors, and misunderstandings about gravity and linear equations across grade levels and reporting patterns in these across countries and by gender. The results reinforce the importance of identifying and understanding students' misconceptions, errors, and misunderstandings to determine what changes may be needed in the curricula through secondary school to improve student learning and to ensure their readiness for post-secondary education and/or future careers.

References

Adu, E., Assuah, C. K., & Asiedu-Addo, S. K. (2015). Students' errors in solving linear equation word problems: Case study of a Ghanaian senior high school. *African Journal of Educational Studies in Mathematics and Sciences, 11,* 17–30.

Angell, C. (2004). *Exploring students' intuitive ideas based on physics items in TIMSS-1995*. In C. Papanastasiou (Ed.), *Proceedings of the IRC-2004 TIMSS. IEA International Research Conference* (Vol. 2, pp. 108–123). Published Nicosia, Cyprus: University of Cyprus. Retrieved from https://www.iea.nl/sites/default/files/2019-03/IRC2004_Angell.pdf.

Bar, V., Brosh, Y., & Sneider, C. (2016). Weight, mass, and gravity: Threshold concepts in learning science. *Science Educator, 25*(1), 22–34.

Bishop, A., Filloy, E., & Puig, L. (2008). *Educational algebra: A theoretical and empirical approach*. Boston, MA, USA: Springer.

Bradshaw, L., & Templin, J. (2014). Combining item response theory and diagnostic classification models: A psychometric model for scaling. *Psychometrika, 79*(3), 403–425.

Caglayan, G., & Olive, J. (2010). Eighth grade students' representations of linear equations based on a cups and tiles model. *Educational Studies in Mathematics, 74*(2), 143–162.

Cavallo, A. M., Potter, W. H., & Rozman, M. (2004). Gender differences in learning constructs, shifts in learning constructs, and their relationship to course achievement in a structured inquiry, yearlong college physics course for life science majors. *School Science and Mathematics, 104,* 288–300.

Choi, K. M., Lee, Y. S., & Park, Y. S. (2015). What CDM can tell about what students have learned: An analysis of TIMSS eighth grade mathematics. *Eurasia Journal of Mathematics, Science and Technology Education, 11*(6), 1563–1577.

Clement, J. (1982). Students' preconceptions in introductory mechanics. *American Journal of Physics, 50*(1), 66–71.

Darling, G. (2012). How does force affect motion? *Science and Children, 50*(2), 50–53.

De la Torre, J., & Minchen, N. (2014). Cognitively diagnostic assessments and the cognitive diagnosis model framework. *Psicología Educativa, 20*(2), 89–97.

Demirci, N. (2005). A study about students' misconceptions in force and motion concepts by incorporating a web-assisted physics program. *Turkish Online Journal of Educational Technology, 4*(3), 40–48.

Docktor, J., & Heller, K. (2008). Gender differences in both force concept inventory and introductory physics performance. *AIP Conference Proceedings, 1064,* 15–18. Retrieved from https://doi.org/10.1063/1.3021243.

Dogan, E., & Tatsuoka, K. (2008). An international comparison using a diagnostic testing model: Turkish students' profile of mathematical skills on TIMSS-R. *Educational Studies in Mathematics, 68*(3), 263–272.

Dyke, F. V., & White, A. (2004). Examining students' reluctance to use graphs. *Mathematics Teacher, 98*(2), 110–117.

Eryilmaz, A. (2002). Effects of conceptual assignments and conceptual change discussions on students' misconceptions and achievement regarding force and motion. *Journal of Research in Science Teaching, 39*(1), 1001–1015.

Gilmore, C., Keeble, S., Richardson, S., & Cragg, L. (2017). The interaction of procedural skill, conceptual understanding and working memory in early mathematics achievement. *Journal of Numerical Cognition, 3*(2), 400–416.

Gönen, S. (2008). A study on student teachers' misconceptions and scientifically acceptable conceptions about mass and gravity. *Journal of Science Education and Technology, 17*(1), 70–81.

Hake, R. R. (2002). Relationship of individual student normalized learning gains in mechanics with gender, high-school physics, and pretest scores on mathematics and spatial visualization. *Physics Education Research Conference*, Boise, Idaho, USA, August 2002. Retrieved from https://www.researchgate.net/publication/237457456_Relationship_of_Individual_Student_Normalized_Learning_Gains_in_Mechanics_with_Gender_High-School_Physics_and_Pretest_Scores_on_Mathematics_and_Spatial_Visualization.

Halloun, I. A., & Hestenes, D. (1985a). Common sense concepts about motion. *American Journal of Physics, 53*(11), 1056–1065.

Halloun, I. A., & Hestenes, D. (1985b). The initial knowledge state of college physics students. *American Journal of Physics, 53*(11), 1043–1048.

Hazari, Z., Tai, R. H., & Sadler, P. M. (2007). Gender differences in introductory university physics performance: The influence of high school physics preparation and affective factors. *Science Education, 91*, 847–876. Retrieved from https://onlinelibrary.wiley.com/doi/abs/10.1002/sce.20223.

Henson, A. H., Templin, J. L., & Willse, J. T. (2009). Defining a family of cognitive diagnosis models using log-linear models with latent variables. *Psychometrika, 74*(2), 191–210.

Hestenes, D., Wells, M., & Swackhamer, G. (1992). Force concept inventory. *The Physics Teacher, 30*(3), 141–158.

Hsu, C. L., Wang, W. C., & Chen, S. Y. (2013). Variable-length computerized adaptive testing based on cognitive diagnosis models. *Applied Psychological Measurement, 37*(7), 563–582.

Huebner, A., & Wang, C. (2011). A note on comparing examinee classification methods for cognitive diagnosis models. *Educational and Psychological Measurement, 71*(2), 407–419.

Huntley, M. A., Marcus, R., Kahan, J., & Miller, J. L. (2007). Investigating high-school students' reasoning strategies when they solve linear equations. *Journal of Mathematical Behavior, 26*, 115–139.

Juan, A., Hannan, S., Zulu, N., Harvey, J. C., Prinsloo, C. H., Mosimege, M., & Beku, U. (2017). *TIMSS item diagnostic report: South Africa: Grade 5 Numeracy. (Commissioned by the Department of Basic Education)*. South Africa: Human Sciences Research Council. Retrieved from http://ecommons.hsrc.ac.za/handle/20.500.11910/11447.

Junker, B. W., & Sijtsma, K. (2001). Cognitive assessment models with few assumptions, and connections with nonparametric item response theory. *Applied Psychological Measurement, 25*(3), 258–272.

Kalchman, M., & Koedinger, K. R. (2005). Teaching and learning functions. In M. S. Donovan, & J. D. Bransford (Eds.), *How students learn: History, mathematics, and science in the classroom* (pp. 351–393). Washington, DC: National Academies Press.

Kavanagh, C., & Sneider, C. (2007). Learning about gravity I. Free fall: A guide for teachers and curriculum developers. *Astronomy Education Review, 5*(21), 21–52.

Kilpatrick, J., & Izsák, A. (2008). A history of algebra in the school curriculum. *Algebra and Algebraic Thinking in School Mathematics, 70*, 3–18.

Kim, E., & Pak, S. J. (2002). Students do not overcome conceptual difficulties after solving 1000 traditional problems. *American Journal of Physics, 70*(7), 759–765.

Knuth, E. J. (2000). Student understanding of the Cartesian connection: An exploratory study. *Journal for Research in Mathematics Education, 31*(4), 500–508.

Kost, L. E., Pollock, S. J., & Finkelstein, N. D. (2009). Unpacking gender differences in students' perceived experiences in introductory physics. *AIP Conference Proceedings, 1179*, 177–180. Retrieved from https://doi.org/10.1063/1.3266708.

Leighton, J. P., & Gierl, M. J. (2007). Defining and evaluating models of cognition used in educational measurement to make inferences about examinees' thinking processes. *Educational Measurement: Issues and Practice, 26*(2), 3–16.

Maris, E. (1999). Estimating multiple classification latent class models. *Psychometrika, 64*, 187–212.

Mosimege, M., Beku, U., Juan, A., Hannan, S., Prinsloo, C. H., Harvey, J. C., & Zulu, N. (2017). *TIMSS item diagnostic report: South Africa: Grade 9 Mathematics. (Commissioned by the Department of Basic Education)*. South Africa: Human Sciences Research Council. Retrieved from http://repository.hsrc.ac.za/handle/20.500.11910/11448.

NCTM. (1989). *Curriculum and evaluation standards for school mathematics*. Reston, VA: National Council of Teachers of Mathematics.

Nussbaum, J. (1979). Children's conceptions of the earth as a cosmic body: A cross age study. *Science Education, 63*(1), 83–93.

Pablico, J. R. (2010). *Misconceptions on force and gravity among high school students*. Louisiana State University Master's theses 2462. Retrieved from https://digitalcommons.lsu.edu/gradschool_theses/2462/.

Palmer, D. (2001). Students' alternative conceptions and scientifically acceptable conceptions about gravity. *International Journal of Science Education, 23*(7), 691–706.

Piburn, M. D., Baker, D. R., & Treagust, D. F. (1988). *Misconceptions about gravity held by college students*. Paper presented at the annual meeting of the National Association for Research in Science Teaching, Lake of the Ozarks, MO, USA, April 10–13, 1988. Retrieved from https://eric.ed.gov/?id=ED292616.

Powell, A. N. (2013). *A study of middle school and college students' misconceptions about solving multi-step linear equations*. Masters thesis, State University of New York at Fredonia, NY, USA. Retrieved from http://hdl.handle.net/1951/58371.

Prinsloo, C. H., Harvey, J. C., Mosimege, M., Beku, U., Juan, A., Hannan, S., & Zulu, N. (2017). *TIMSS item diagnostic report: South Africa: Grade 9 Science. (Commissioned by the Department of Basic Education)*. South Africa: Human Sciences Research Council. Retrieved from http://www.hsrc.ac.za/en/research-data/ktree-doc/19286.

Provasnik, S., Malley, L., Neidorf, T, Arora, A., Stephens, M., Balestreri, K., Perkins, R., & Tang, J. H. (2019, in press). *U.S. performance on the 2015 TIMSS Advanced mathematics and physics assessments: A closer look (NCES 2017-020)*. Washington, DC: US Department of Education, National Center for Education Statistics.

Reif, F., & Allen, S. (1992). Cognition for interpreting scientific concepts: A study of acceleration. *Cognition and Instruction, 9*(1), 1–44.

Roussos, L. A., Templin, J. L., & Henson, R. A. (2007). Skills diagnosis using IRT-based latent class models. *Journal of Educational Measurement, 44*(4), 293–311.

Saputro, B. A., Suryadi, D., Rosjanuardi, R., & Kartasasmita, B. G. (2018). Analysis of students' errors in responding to TIMSS domain algebra problem. *Journal of Physics: Conference Series, 1088*, 012031. Retrieved from https://iopscience.iop.org/article/10.1088/1742-6596/1088/1/012031.

Shear, B. R., & Roussos, L. A. (2017). Validating a distractor-driven geometry test using a generalized diagnostic classification model. In B. Zumbo, & A. Hubley (Eds.), *Understanding and investigating response processes in validation research* (Vol. 69, pp. 277–304). Cham, Switzerland: Springer.

Simon, M. A., & Blume, G. W. (1994). Mathematical modeling as a component of understanding ratio-as-measure: A study of prospective elementary teachers. *Journal of Mathematical Behavior, 13*, 183–197.

Sneider, G., & Pulos, S. (1983). Children's cosmographies: Understanding the earth's shape and gravity. *Science Education, 67*(2), 205–221.

Stein, M., Larrabee, T. G., & Barman, C. R. (2008). A study of common beliefs and misconceptions in physical science. *Journal of Elementary Science Education, 20*(2), 1–11.

Stump, S. L. (2001). High school precalculus students' understanding of slope as measure. *School Science and Mathematics, 101*(2), 81–89.

Swafford, J. O., & Langrall, C. W. (2000). Grade 6 students' preinstructional use of equations to describe and represent problem situations. *Journal for Research in Mathematics Education, 31* (1), 89–112.

Tatsuoka, K. K. (1983). Rule space: An approach for dealing with misconceptions based on item response theory. *Journal of Educational Measurement, 20*(4), 345–354.

Thornton, R. K., Kuhl, D., Cummings, K., & Marx, J. (2009). Comparing the force and motion conceptual evaluation and the force concept inventory. *Physical Review Special Topics: Physics Education Research, 5*(1), 1–8.

Thornton, R. K., & Sokoloff, D. R. (1998). Assessing student learning of Newton's laws: The force and motion conceptual evaluation and the evaluation of active learning laboratory and lecture curricula. *American Journal of Physics, 66*(4), 338–352.

Trowbridge, D. E., & McDermott, L. C. (1980). Investigation of student understanding of the concept of velocity in one dimension. *American Journal of Physics, 48*(12), 1020–1028.

Tsamir, P., & Almog, N. (2001). Students' strategies and difficulties: the case of algebraic inequalities. *International Journal of Mathematical Education in Science and Technology, 32* (4), 513–524.

Văcăreţu, A. (n.d.). *Using the TIMSS results for improving mathematics learning.* Cluj-Napoca, Romania: Romanian Reading and Writing for Critical Thinking Association. Retrieved from http://directorymathsed.net/montenegro/Vacaretu.pdf.

Yamaguchi, K., & Okada, K. (2018). Comparison among cognitive diagnostic models for the TIMSS 2007 fourth grade mathematics assessment. *PLoS ONE, 13*(2), e0188691.

Yung, B. H. W. (Ed.). (2006). *Learning from TIMSS: Implications for teaching and learning science at the junior secondary level. TIMSS HK IEA Centre.* Hong Kong: Education and Manpower Bureau. Retrieved from https://cd1.edb.hkedcity.net/cd/science/is/timss/learningfromtimss.pdf.

Chapter 3
Methodology Used to Analyze Student Misconceptions, Errors, and Misunderstandings in TIMSS

Abstract The Trends in International Mathematics and Science Study (TIMSS) and TIMSS Advanced assessments are a good source of data for the study of student misconceptions, errors, and misunderstandings in physics and mathematics. After examining the available range of TIMSS and TIMSS Advanced data, five countries that participated in the TIMSS 2015 and TIMSS Advanced 2015 assessments, and all, or most, of the prior TIMSS assessments, were selected for study (Italy, Norway, Slovenia, the Russian Federation, and the United States) to maximize the cross-country comparisons that could be made across grade levels and assessment years. A complete review of the TIMSS and TIMSS Advanced assessment frameworks and content mapping (to determine related topics and items across grades and assessment cycles) identified the set of items that measure misconceptions, errors, and misunderstandings related to the topics of gravity and linear equations. Item-level statistics (the percentage of students who provided the correct answer, and the percentage demonstrating the misconception, error, or misunderstanding) were used to make comparisons across countries at each grade level overall and by gender. In addition to analyzing gender differences, examining trends in patterns of misconceptions, errors, and misunderstandings over time provided important information across countries.

Keywords Diagnostic data · Errors · International large-scale assessment · Item statistics · Misconceptions · Student achievement · *t*-test · Trend analysis · Trends in International Mathematics and Science Study (TIMSS) · Italy · Norway · Russian Federation · Slovenia · United States

3.1 TIMSS and TIMSS Advanced Data

TIMSS and TIMSS Advanced assessments have been measuring trends in international mathematics and science achievement since 1995, based on nationally representative samples of students in each participating country at grade four, grade eight, and the final year of secondary school (for students taking advanced

T. Neidorf et al., *Student Misconceptions and Errors in Physics and Mathematics*, IEA Research for Education 9, https://doi.org/10.1007/978-3-030-30188-0_3

coursework in physics and mathematics). TIMSS has been administered every four years for six assessment cycles[1] (namely in 1995, 1999, 2003, 2007, 2011, and 2015), while TIMSS Advanced has been administered at three points in time (1995, 2008, and 2015). Following the release of the international reports from each assessment, the IEA releases international databases for secondary analyses. In addition, after each assessment, a portion of the assessment items (and scoring guides) are released, while at least half are retained as secure items for future assessment cycles. In both assessments, items may be released after one, two, or three assessment cycles.

This report used assessment items and student performance data from the TIMSS and TIMSS Advanced assessments conducted across all assessment cycles from 1995 to 2015. The set of countries administering TIMSS and TIMSS Advanced varies for each assessment cycle. We report on five countries that participated in the TIMSS Advanced 2015 assessment and in all, or most, of the TIMSS grade eight and grade four mathematics and science assessments since 1995: Italy, Norway, the Russian Federation, Slovenia, and the United States (see Tables 3.1, 3.2, and 3.3). These countries were selected from the nine countries participating in TIMSS Advanced 2015 (Table 3.1) to maximize the data available to answer the research questions (see Sect. 1.4). All selected countries participated at all three grade levels

Table 3.1 Participation of countries in TIMSS Advanced assessments, by cycle

Country	TIMSS Advanced					
	Advanced mathematics			Physics		
	1995	2008	2015	1995	2008	2015
France	•	–	•	•	–	•
Italy	•	•	•	–	•	•
Lebanon	–	•	•	–	•	•
Norway	–	•	•	•	•	•
Portugal	–	–	•	–	–	•
Russian Federation	•	•	•	•	•	•
Slovenia	•	•	•	•	•	•
Sweden	•	•	•	•	•	•
United States	•	–	•	•	–	•

• Indicates participation in that assessment cycle
– Indicates no participation
Notes This table includes the nine countries that participated in the TIMSS Advanced 2015 assessment. Five countries were selected for inclusion in this study (Italy, Norway, the Russian Federation, Slovenia, and the United States). See Appendix for TIMSS Advanced data considerations for 1995, 2008, and 2015
Source TIMSS Advanced 1995, 2008, and 2015 assessments. Copyright International Association for the Evaluation of Educational Achievement (IEA). Publisher: TIMSS & PIRLS International Study Center, Lynch School of Education, Boston College. Retrieved from Mullis et al. (2016a, Appendix MA.1 and Appendix PA.1)

[1]In 1999, the TIMSS assessment was only administered at grade eight.

Table 3.2 Participation of countries in TIMSS grade eight assessments, by cycle

Country	TIMSS grade 8					
	1995	1999	2003	2007	2011	2015
France	•	–	–	–	–	–
Italy	○	•	•	•	•	•
Lebanon	–	–	•	•	•	•
Norway	•	–	•	•	•	•
Portugal	•	–	–	–	–	–
Russian Federation	•	•	•	•	•	•
Slovenia	•	○	•	•	•	•
Sweden	•	–	•	•	•	•
United States	•	•	•	•	•	•

• Indicates participation in that assessment cycle
○ Indicates participation but data not comparable for measuring trends to 2015
– Indicates no participation
Notes This table includes the nine countries that participated in the TIMSS Advanced 2015 assessment. Five countries were selected for inclusion in this study (Italy, Norway, the Russian Federation, Slovenia, and the United States). See Appendix for TIMSS grade eight data considerations for 1995, 1999, and 2015
Source TIMSS 1995, 1999, 2003, 2007, 2011, and 2015 assessments. Copyright International Association for the Evaluation of Educational Achievement (IEA). Publisher: TIMSS & PIRLS International Study Center, Lynch School of Education, Boston College. Retrieved from Mullis et al. (2016b, Appendix A.1)

in the 2015 assessments and were missing data for no more than one assessment cycle at any grade level. The five selected countries thus permit the greatest number of comparisons across countries, grade levels, and assessment cycles.

TIMSS assesses mathematics and science achievement at two grade levels and so has two target populations: all students enrolled in grade four and all students enrolled in grade eight (or the equivalent grades in each country). The TIMSS Advanced physics and mathematics populations are defined as students in their final year of secondary school who are currently taking (or who had previously taken) the TIMSS Advanced-eligible courses in physics or advanced mathematics[2] (Martin et al. 2014). (More information is provided about the TIMSS and TIMSS Advanced populations in Appendix.)

The TIMSS Advanced population is a select group reflecting one-quarter or less of final-year students in most countries in 2015. The coverage index (percentage of the corresponding age cohort covered by the TIMSS Advanced physics and advanced mathematics student populations) was lower in physics than in advanced mathematics in all five countries included in the study (Tables 3.4 and 3.5). For physics, the coverage index in 2015 ranged from about 5% coverage in the Russian Federation and the United States, to 18% in Italy (Table 3.4). There were some

[2]TIMSS Advanced-eligible courses are defined as those that cover most of the topics outlined in the TIMSS Advanced physics and mathematics assessment frameworks.

Table 3.3 Participation of countries in TIMSS grade four assessments, by cycle

Country	TIMSS grade 4				
	1995	2003	2007	2011	2015
France	–	–	–	–	•
Italy	○	•	•	•	•
Lebanon	–	–	–	–	–
Norway	•	•	•	•	•
Portugal	•	–	–	•	•
Russian Federation	–	•	•	•	•
Slovenia	•	•	•	•	•
Sweden	–	–	•	•	•
United States	•	•	•	•	•

• Indicates participation in that assessment cycle
○ Indicates participation but data not comparable for measuring trends to 2015
– Indicates no participation
Notes This table includes the nine countries that participated in the TIMSS Advanced 2015 assessment. Five countries were selected for inclusion in this study (Italy, Norway, the Russian Federation, Slovenia, and the United States). See Appendix for TIMSS grade four data considerations for 1995 and 2015. TIMSS was not administered at grade four in 1999
Source TIMSS 1995, 2003, 2007, 2011, and 2015 assessments. Copyright International Association for the Evaluation of Educational Achievement (IEA). Publisher: TIMSS & PIRLS International Study Center, Lynch School of Education, Boston College. Retrieved from Mullis et al. (2016b, Appendix A.1)

differences in the physics coverage index across the assessment years, with increases seen in Italy, the Russian Federation, and the United States, and decreases seen in Norway and Slovenia between 2015 and 2008 or 1995. In particular, the percentage of students studying physics at an advanced level in Italy increased from 4% in 2008 to 18% in 2015, which indicates a progressively more inclusive sample of students. In contrast, the percentage of students decreased in Slovenia from 39% in 1995 to 7–8% in 2008 and 2015, reflecting a more restricted sample of students. For advanced mathematics, the coverage index in 2015 ranged from 10 to 11% in the Russian Federation and the United States, to 34% in Slovenia (Table 3.5; see Appendix for any additional data considerations between 1995 and 2015).

In addition to the overall coverage index, the percentages of female and male students in the TIMSS Advanced populations (final-year students taking advanced coursework in physics or mathematics) varied across countries and may differ from the percentages in the full population of students in their final year of secondary school. Boys were more likely to undertake advanced physics coursework than girls in all five countries (Table 3.4); only about 30% of advanced physics students in Norway and Slovenia, and about 40% of students in Italy, the Russian Federation, and the United States were female. The percentage of females in physics did not change substantially across assessment years in any country. In contrast to physics, the percentage of female students taking advanced mathematics (Table 3.5) was lower than males in Italy and Norway (about 40%), higher than males in Slovenia

Table 3.4 Coverage index and percentages of female and male students in TIMSS Advanced physics samples, by country and year

Country	Coverage index (%)[1]			Percentage of students[2]					
				Female			Male		
	2015	2008	1995	2015	2008	1995	2015	2008	1995
Italy	18.2	3.8	–	46 (1.1)	40 (2.4)	–	54 (1.1)	60 (2.4)	–
Norway	6.5	6.8	8.4	29 (1.2)	29 (1.7)	26 (1.8)	71 (1.2)	71 (1.7)	74 (1.8)
Russian Federation	4.9	2.6	1.5	42 (1.2)	45 (1.3)	46 (2.0)	58 (1.2)	55 (1.3)	54 (2.0)
Slovenia	7.6	7.5	38.6	30 (1.7)	27 (1.2)	28 (3.7)	70 (1.7)	73 (1.2)	72 (3.7)
United States	4.8	–	2.7	39 (1.6)	–	43 (4.7)	61 (1.6)	–	57 (4.7)

Notes

– Indicates no participation

[1] The physics coverage index is the percentage of the corresponding age cohort covered by the TIMSS Advanced target population (students in their final year of secondary school who were taking or had previously taken TIMSS Advanced-eligible physics courses)

[2] Standard errors are provided in brackets

Source TIMSS Advanced 1995, 2008, and 2015 assessments. Copyright International Association for the Evaluation of Educational Achievement (IEA). Publisher: TIMSS & PIRLS International Study Center, Lynch School of Education, Boston College. Retrieved from Mullis et al. (2016a, Exhibits P1.4 and P1.7)

Table 3.5 Coverage index and percentages of female and male students in TIMSS Advanced 2015 mathematics samples, by country

Country	Coverage index (%)[1]	Percentage of students[2]	
		Female	Male
Italy	24.5	37 (1.3)	63 (1.3)
Norway	10.6	38 (1.4)	62 (1.4)
Russian Federation	10.1	50 (1.3)	50 (1.3)
Slovenia	34.4	60 (1.1)	40 (1.1)
United States	11.4	49 (0.9)	51 (0.9)

Notes Data for 2015 only, since there were no TIMSS Advanced mathematics items related linear equations from TIMSS 1995 or TIMSS 2008 included in the study

[1]The mathematics coverage index is the percentage of the corresponding age cohort covered by the TIMSS Advanced target population (students in their final year of secondary school who were taking or had previously taken TIMSS Advanced-eligible mathematics courses)

[2]Standard errors are provided in brackets. Because of rounding some results may appear inconsistent

Source TIMSS Advanced 2015 assessment. Copyright International Association for the Evaluation of Educational Achievement (IEA). Publisher: TIMSS & PIRLS International Study Center, Lynch School of Education, Boston College. Retrieved from Mullis et al. (2016a, Exhibits M1.4 and M1.7)

(about 60%), and about equal to males in the Russian Federation and the United States.

All results in this report are based on item-level statistics available using the TIMSS and TIMSS Advanced international databases from each assessment cycle, including the weighted percent correct for each country and the percentage of students in each item response category (see Sect. 3.2.3). Item-level statistics were computed for each country, as well as on average across the five countries included in the study (overall and by gender). Example items used in this report include "restricted-use" items[3] from the TIMSS 2015 assessments, as well as released items from prior assessment years. Although all example items are released or restricted-use items, appropriate non-released (secure) items from TIMSS 2015 were included in the analyses of patterns in misconceptions, but are not shown in the report.

[3]The 2015 "restricted-use" items are those designated by the TIMSS & PIRLS International Study Center for use as examples in the international report as well as by participating countries in their national reports or for research purposes, such as this IEA thematic report. Example items from 2015 included in this report are used with permission from the IEA. Secure items from 2015 are discussed but are not shown in the report.

3.2 Methodology

Our methodology consisted of three major components: (1) assessment framework review and content mapping to identify the set of items measuring the selected topics in our study (gravity and linear equations); (2) evaluation of diagnostic item-level performance data to identify the specific performance objectives measured by these items and to provide evidence of specific types of misconceptions, errors, and misunderstandings; and (3) analyses of the percentage of students demonstrating these misconceptions, errors, and misunderstanding to report patterns across countries by grade level, gender, and assessment year.

3.2.1 Assessment Framework Review and Content Mapping

To determine how mathematics and science concepts progress from the lower grades in TIMSS to TIMSS Advanced, topics covered in the 2015 TIMSS Advanced assessment frameworks were mapped to related topics at grades four and eight in the TIMSS 2015 frameworks. In the TIMSS and TIMSS Advanced frameworks, the greatest degree of content overlap across grades four, eight, and 12 is in the physics topic area of mechanics (forces and motion) and the mathematics content area of algebra, resulting in adequate numbers of assessment items across grades to report on patterns of misconceptions. Within topics, a set of framework objectives were identified at each grade level that were then used to select the items used in the study.

As described in Chap. 1, this study focuses on two specific topics: gravity in physics and linear equations in algebra. We determined the set of TIMSS 2015 and TIMSS Advanced 2015 framework objectives that measured these topics (or precursor topics) across grade levels for gravity (Table 1.1) and linear equations (Table 1.2). Since the TIMSS and TIMSS Advanced frameworks have been revised over the past 20 years, content mapping also included mapping the TIMSS framework objectives in 1995, 1999, 2003, 2007, and 2011, and the TIMSS Advanced framework objectives in 1995 and 2008, to the corresponding TIMSS 2015 framework objectives.

3.2.2 Evaluation of Item-Level Performance Data

Once the specific TIMSS and TIMSS Advanced framework objectives related to gravity and linear equations were identified, sets of items for each topic (16 for physics and 28 for mathematics) from the grade four, grade eight, and TIMSS Advanced assessments were assembled and reviewed. First, the TIMSS Advanced

2015 items were evaluated to determine the performance objectives measured by each item and the specific types of misconceptions, errors, and misunderstandings demonstrated by students across the five TIMSS Advanced countries chosen for the study (Italy, Norway, the Russian Federation, Slovenia, and the United States).[4] Then, TIMSS items from across the assessment cycles at grades four and eight that measured related or precursor concepts were evaluated for evidence of specific misconceptions, errors, and misunderstandings at the lower grade levels.[5]

Evidence of misconceptions, errors, and misunderstandings was determined by examining patterns in the item-level performance data. For multiple-choice (MC) items, this involved distractor analysis, or examining the incorrect options to determine common errors and misconceptions that may be demonstrated by students who choose those options. For constructed-response (CR) items (where students provide a written response), response patterns were determined based on the nature of student responses as defined in the scoring guides that accompany the items. In TIMSS and TIMSS Advanced, scoring guides provide item-specific criteria to differentiate between correct, partial, and incorrect student responses and use two-digit diagnostic codes to track specific misconceptions or errors (i.e., to differentiate between different types of partial and incorrect responses). This initial item evaluation used item statistics (i.e., the weighted percentage distributions of students in each country choosing each MC response option or each CR item response category in the scoring guide) obtained from the international data almanacs available on the TIMSS & PIRLS International Study Center website (https://timssandpirls.bc.edu/).

Further content analysis of the set of items covering the topics of gravity and linear equations at each grade level identified a set of performance objectives (four in physics and nine in mathematics) that were measured by these items across the grade levels. These performance objectives are based on the set of TIMSS and TIMSS Advanced items selected for the study and are more specific than the broader TIMSS and TIMSS Advanced framework objectives outlined in Chap. 1. Some performance objectives were assessed at only one grade level, while others were measured by items at two grade levels (i.e., TIMSS Advanced/grade eight or grade eight/grade four) or at all three grade levels (for physics only). For the items measuring each performance objective, we identified the misconceptions, errors, or misunderstandings that may be demonstrated by different types of incorrect student responses. There were from one to six items measuring each type of misconception,

[4]Additional TIMSS Advanced items from 1995 and 2008 were also evaluated for physics. Mathematics only included items from TIMSS Advanced 2015.

[5]The TIMSS testing schedule permits the same cohort of students to be assessed over time (e.g., grade four students in 2007 are grade eight students in 2011, and grade 12 students in 2015). This report does not directly measure changes in specific misconceptions, errors, and misunderstandings over time for the same cohort of students due to limitations in the available item-level data. This raises some potential considerations and implications for future research in this area (see Sect. 5.4).

error, and misunderstanding. (See Sect. 1.2 for detailed definitions of the terms, and Chap. 4 for an overview of performance objectives, misconceptions, errors, and misunderstandings, and the set of items used in the study.)

3.2.3 Reporting Patterns in Percent Correct and Percent with Misconceptions, Errors, and Misunderstandings by Grade, Country, Gender, and Assessment Year

All of the analyses used to report on the percent correct and percentage of students with misconceptions, errors, and misunderstandings were conducted using the IEA's International Database (IDB) Analyzer (Version 4.0) *Percentages* function (IEA 2018). The IDB Analyzer uses a jackknife repeated replication (JRR) procedure to compute estimates and standard errors for a variety of statistics, such as average scores and percent correct (see Appendix for further technical notes). We do not provide standard errors in the tables and figures in this book (supplementary materials providing standard errors for all estimates are available for download at www.iea.nl/publications/RfEVol9).

Four types of analyses were used to produce the item-level statistics shown in the report.

Percent Correct
This is the percentage of students receiving credit on each item. For MC and short CR items (each worth one score point), this reflects the percentage of students who provided a correct answer. For extended CR items, this reflects the weighted percentage of students receiving full credit (two points) or partial credit (one point). For example, on an item where 10% of students received full credit and 10% received partial credit, the weighted percent correct is 15%, which reflects the percentage of students receiving full credit (10%) plus half the percentage receiving partial credit (5%). Percent correct was computed for all items in each country (overall and by gender). When reporting percent correct on the set of items in physics and mathematics, data from the most recent assessment was used for each item.

Percentage of Students with Misconceptions, Errors, and Misunderstandings
Two different types of item-level analyses were used to determine these percentages:

(1) Specific types of misconceptions and misunderstandings reflected items where a single response option (in the case of MC items) or a single scoring guide category (in the case of CR items), or multiple response options or multiple scoring guide categories were identified to track and report on a particular type of misconception or misunderstanding. The percentage of students with the specific misconception (in physics) or misunderstanding (in mathematics) was calculated as the sum of the percentages of students in each of the relevant options or score categories. Specific misconceptions and misunderstandings apply to 11 items in physics (10 MC and one CR) and three items in

mathematics (all MC). For two of the MC items in physics, one response option measured one type of misconception and others measured a second type; two separate analyses were conducted to obtain the percentages for both types of misconceptions.

(2) General types of misunderstandings reflected items where there were no specific misconceptions, errors, or misunderstandings tracked. All that could be determined was whether or not a student was able to demonstrate the understanding or ability required for the performance objective measured by the item. For these items, the percentage of students with a more general type of misunderstanding reflected all students who did not answer the item correctly. This included students who attempted the item but provided an incorrect response (including invalid responses or off-task comments), as well as those students who did not answer the item (omitted responses).[6] General types of misunderstandings apply to six items in physics (one MC and five CR) and 26 items in mathematics (12 MC and 14 CR). The majority of these items were constructed response and many required students to explain their answer or show their work. In the TIMSS scoring guides, the general incorrect code 79 covers any type of incorrect response, including "crossed out, erased, stray marks, illegible, or off task" responses. When including blanks (omitted responses), we assumed that students who reached the item, but did not respond, did not have the understanding necessary to answer the question (i.e., similar in nature to responses that contain stray marks or off-task comments). This is consistent with the assumption underlying TIMSS scale scores, where omitted responses are treated as incorrect in scaling. The alternative would be to remove the blanks (omitted responses) from the sample, which would underestimate the percentage of students who did not demonstrate conceptual understanding.

The codes used for specific or general types of misconceptions, errors, and misunderstandings, and the corresponding value labels in the TIMSS data files are provided for all physics and mathematics items in Appendix (Tables A.1 and A.2).[7] The percentage of students with misconceptions, errors, and misunderstandings was computed for all items in each country (overall and by gender). For trend items administered in multiple assessments, the percentage of students was reported for each assessment year.

Average Percent Correct and Average Percent with Misconceptions, Errors, and Misunderstandings

These averages reflect the percent correct (or percent with misconceptions, errors, or misunderstandings) in each country averaged across the countries that have data for the item. For most items, this reflects the average across all five countries.

[6]The "percent omitted" does not include the percent "not reached;" a response that is "not reached" is treated as a missing response and is not included in the denominator for the percent correct or percent with misconceptions.

[7]The separate analyses for the physics items that measured two different types of misconceptions were identified by two different versions (V1 and V2; see Table A.1).

However, there were some assessment years where data were not available for all countries, and the averages were based only on three or four countries.

3.2.4 Statistical Comparisons

Differences in the percent correct and the percentage of students with misconceptions, errors, and misunderstandings were computed (1) between each country and the average across the five countries, (2) between female and male students within each country, and (3) across assessment years for the trend items. The appropriate t-tests were used for all comparisons involving these item-level statistics, and indicators of statistical significance are provided in all data tables and figures that provide comparisons. A difference was considered "significant" when the probability (p) associated with the t-test was less than 0.05 (i.e., the probability is at least 95% that the reported difference is "real" and not due to chance). We used the following t-tests for each type of comparison.

(1) For comparisons between the percentages in each country and the average across the five countries, there is overlap between the samples (i.e., each country is part of the average). In such cases, a part-whole t-test was used to account for this overlap:

$$t = \frac{(est_j - est_i)}{\sqrt{se_i^2 + (1 - 2p)se_j^2}}$$

Where est_i is the estimated average percentage for the five countries; est_j is the estimated percentage for one country; se_i and se_j are the respective corresponding standard errors; and p is the proportion of the five countries represented by each country (0.2).

(2) For within-country gender differences, there are two types of t-tests that can be used depending on the student samples: independent (when there are independent random samples of female and male students drawn from the population) and non-independent (when this is not the case).
For independent random samples, the independent t-test is appropriate:

$$t = \frac{(est_{female} - est_{male})}{\sqrt{(se_{female})^2 + (se_{male})^2}}$$

where est_{female} and est_{male} are the estimates for the percentage of females and males, respectively, and se_{female} and se_{male} are the corresponding standard errors of these percentages. The independent t-test can be calculated using the output from the IDB Analyzer, where the JRR procedure is used to determine the separate percentages and standard errors for females and males.

However, in the TIMSS and TIMSS Advanced assessments, the samples of female and male students are not independent, since they are in the same schools and classrooms selected to take the assessments. Therefore, the correct *t*-test for non-independent samples requires the standard error of the difference between the percentage of females and the percentage of males:

$$t = \frac{\left(est_{female} - est_{male}\right)}{se\left(est_{female} - est_{male}\right)}$$

The standard error of the difference, $se\ (est_{female} - est_{male})$, takes into account the covariance (*cov*) between females and males for dependent samples:

$$se_{\left(est_{female} - est_{male}\right)} = \sqrt{se^2_{\left(est_{female}\right)} + se^2_{\left(est_{male}\right)} - 2cov\left(est_{female}, est_{male}\right)}$$

To obtain the appropriate standard errors, the JRR procedure must be conducted on the female–male percentage difference. The version of the IDB Analyzer that we used (Version 4.0), employs the JRR procedure to obtain standard errors for the percent of females and the percent of males. It does not, however, allow for jackknifing the gender differences for these item-level statistics (percent correct or percent misconceptions, errors, and misunderstandings). Therefore, the independent standard errors (and computed *t*-tests) obtained using the IDB Analyzer are approximations that do not take into account the covariance between females and males. These approximations are acceptably accurate if the covariances are small in comparison to the standard errors of the percentage of females and percentage of males. For the item-level statistics, this is expected to be the case due to the design of TIMSS, where only a small number of students take each item. Generally, about four students in each school or class will take each item (Martin et al. 2014).

To determine the magnitude of the covariances for gender differences, we conducted analyses for selected items using the Ed*Survey* R Package[8] (NCES [National Center for Education Statistics] 2018) *Gap* function. The *Gap* function applies the JRR technique to the difference between the percentage of females and males. The output includes the standard error of the difference and the covariance. In the tested cases, we found that the covariance for the

[8]Ed*Survey* is an R statistical package developed by American Institutes for Research (AIR) and commissioned by the National Center for Education Statistics (NCES). Ed*Survey* is tailored to the processing and analysis of NCES large-scale education data with appropriate procedures. Ed*Survey* Version 2.0.3 is designed for the analysis of national and international education data from the NCES, including TIMSS and TIMSS Advanced. For more information, see: https://nces.ed.gov/nationsreportcard/researchcenter/software.aspx

item-level statistics was very small. We then ran analyses on the same items using the IDB Analyzer and compared the standard errors and *t*-tests obtained for gender differences using the two different methods. The standard errors using the correct non-independent method (Ed*Survey* R Package) and those using the independent method (IDB Analyzer) were approximately the same (to the nearest at the 0.0001%), and the significance of reported differences was not affected.[9] Thus, for convenience, we used the output from the IDB Analyzer for the gender differences and applied the approximate independent *t*-tests for all items. Additional information on both software packages, as well as example outputs, are provided in Appendix.

(3) The differences between years for trend items are based on independent samples. Thus, the standard independent *t*-test was used:

$$t = \frac{\left(est_{year1} - est_{year2}\right)}{\sqrt{\left(se_{year1}\right)^2 + \left(se_{year2}\right)^2}}$$

where est_{year1} and est_{year2} are the estimates for the percentage of students in the two assessment years being compared, and se_{year1} and se_{year2} are the corresponding standard errors.

3.3 Addressing the Research Questions

As described in Sect. 3.2.2, we reviewed the set of TIMSS and TIMSS Advanced items that measured student understanding of the key concepts (gravity in physics and linear equations in mathematics), administered in each assessment year from 1995 to 2015. As a consequence, we established performance objectives that could be assessed by the items across grade levels and the types of student misconceptions, errors, and misunderstandings demonstrated on these items. This enabled us to report student performance on the TIMSS and TIMSS Advanced items related to gravity and linear equations across countries by grade level, gender, and assessment year to answer the three research questions.

[9]See Appendix (Tables A.3 and A.4) for comparison of the output from the IDB Analyzer and Ed*Survey* R package.

3.3.1 Research Question 1

What are common types of student misconceptions, errors, and misunderstandings in grade four, grade eight, and the final year of secondary school (TIMSS Advanced students), and how do they compare across countries?

To answer the first research question, we examined items from TIMSS and TIMSS Advanced administered at each grade level that demonstrated specific types of student misconceptions, misunderstandings, and errors. We determined the percentage of students for each response type by country and on average across the five countries included in the study. Response patterns provided evidence of the nature and extent of students' misconceptions, misunderstandings, and errors. We presented released example items at each grade level to illustrate the different types of misconceptions, errors, and misunderstandings. Each example item exhibit showed the item; the scoring guide (for CR items) or correct answer (for MC items); and other item information, including the TIMSS item ID,[10] year(s) administered, and the performance objective assessed by the item.[11]

3.3.2 Research Question 2

How do student misconceptions, errors, and misunderstandings differ by gender?

To answer the second research question, we determined the percentage of male students and percentage of female students demonstrating each type of misconception, misunderstanding, and error. We prepared tree graphs showing the gender differences across countries at each grade level.

3.3.3 Research Question 3

How persistent are patterns in misconceptions, errors, and misunderstandings over time?

To answer the third research question, we plotted figures for each country showing the percentage of students demonstrating the specific types of misconceptions, misunderstandings, and errors over multiple assessment years based on the set of trend items.

[10]The item IDs are those used in the TIMSS and TIMSS Advanced databases and released item sets, allowing readers to access all the released items used in this report.

[11]The performance objectives are those developed for this report. These generally are more detailed than the broader TIMSS and TIMSS Advanced framework objectives and reflect the specific set of items included.

References

IEA. (2018). *IDB Analyzer (Version 4.0)*. Hamburg, Germany: IEA. Retrieved from http://www.iea.nl/data.

Martin, M. O., Mullis I. V. S., & Foy, P. (2014). TIMSS Advanced 2015 assessment design. In I. V. S. Mullis, & M. O. Martin (Eds.), *TIMSS Advanced 2015 assessment frameworks* (pp. 85–98). Chestnut Hill, MA: TIMSS & PIRLS International Study Center, Boston College. Retrieved from: https://timssandpirls.bc.edu/timss2015-advanced/frameworks.html.

Mullis, I. V. S., Martin, M. O., Foy, P., & Hooper, M. (2016a). *TIMSS Advanced 2015 international results in advanced mathematics and physics*. Chestnut Hill, MA: TIMSS & PIRLS International Study Center, Boston College. Retrieved from http://timssandpirls.bc.edu/timss2015/international-results/advanced/.

Mullis, I. V. S., Martin, M. O., Foy, P., & Hooper, M. (2016b). *TIMSS 2015 international results in mathematics*. Chestnut Hill, MA: TIMSS & PIRLS International Study Center, Boston College. Retrieved from http://timssandpirls.bc.edu/timss2015/international-results/.

NCES. (2018). Ed*Survey (Version 2.0.3)*. *US Department of Education*. Washington, DC: National Center for Education Statistics. Retrieved from https://cran.r-project.org/package=EdSurvey.

Chapter 4
Results for Student Misconceptions, Errors, and Misunderstandings in Physics and Mathematics

Abstract Diagnostic item-level student performance data from twenty years of TIMSS and TIMSS Advanced assessments can be used to explore students' level of understanding of gravity and linear equations across grades four, eight, and the final year of secondary school (TIMSS Advanced students). Sets of assessment items at each grade level illustrate the nature and extent of student misconceptions, errors, and misunderstandings across grade levels in five countries (Italy, Norway, the Russian Federation, Slovenia, and the United States). The results include assessment of how students in each country performed on the set of items measuring understanding of the physics and mathematics concepts explored in this study (gravity and linear equations); common types of student misconceptions, errors, and misconceptions across grade levels in each country; patterns in misconceptions, errors, and misunderstandings across countries; and gender differences in the frequency of misconceptions, errors, and misunderstandings in each country. The frequency of specific types of student misconceptions, errors, and misunderstandings at each grade level varied across the five countries. Gender differences were found at all three grade levels, but were more extensive for physics than mathematics. Trend items administered in multiple assessment years indicated that the frequency of certain student misconceptions, errors, and misunderstandings decreased over time, while the frequency of others increased.

Keywords Diagnostic data · Errors · Gender differences · Gravity · International large-scale assessment · Item-level data · Linear equations · Mathematics · Misconceptions · Performance objectives · Physics · Science · Student achievement · Trend analysis · Trends in International Mathematics and Science Study (TIMSS) · Italy · Norway · Russian Federation · Slovenia · United States

T. Neidorf et al., *Student Misconceptions and Errors in Physics and Mathematics*, IEA Research for Education 9, https://doi.org/10.1007/978-3-030-30188-0_4

4.1 Introduction

The results for physics (Sect. 4.2) and mathematics (Sect. 4.3) start with an overview of the set of TIMSS and TIMSS Advanced items that measure student understanding of the key concepts that are the focus of this study (gravity in physics and linear equations in mathematics). The overview also describes the performance objectives that are assessed by the items across grade levels and the types of student misconceptions, errors, and misunderstandings demonstrated on these items.[1]

Following the overview, physics and mathematics results are presented in six subsections that report student performance on the TIMSS and TIMSS Advanced items by grade level, country, gender, and assessment year (see textbox).

Subsections in physics and mathematics results	
4.2.1 (physics) 4.3.1 (mathematics)	How did students perform on TIMSS and TIMSS Advanced items related to the topics of interest? (gravity in physics and linear equations in mathematics)
4.2.2 (physics) 4.3.2 (mathematics)	What are common types of misconceptions, errors, and misunderstandings related to these topics in the final year of secondary school, grade 8, and grade 4, and how do they compare across countries? (shown in example items)
4.2.3 (physics) 4.3.3 (mathematics)	What are the patterns in misconceptions, errors, and misunderstandings across grade levels in each country?
4.2.4 (physics) 4.3.4 (mathematics)	How do student misconceptions, errors, and misunderstandings differ by gender?
4.2.5 (physics) 4.3.5 (mathematics)	How persistent are patterns in misconceptions, errors, and misunderstandings over time?
4.2.6 (physics) 4.3.6 (mathematics)	Summary of results*

*In addition to the separate physics and mathematics sections here, a summary of results across physics and mathematics is provided in Sect. 5.1

We present released example items from TIMSS and TIMSS Advanced to demonstrate specific types of student misconceptions, errors, and misunderstandings along with tabular data showing the percentage of students for each response

[1]See Sect. 1.2 for definitions of the terminology used throughout the report (performance objectives, misconceptions, errors, and misunderstandings) and how these relate to the physics and mathematics items.

type by country and on average across the five countries. All example items shown in the report are the standard international version.[2]

In addition to example items, the results include tables and figures that show patterns in the percentage of students demonstrating specific misconceptions, errors, and misunderstandings across countries and grades based on the set of items that measure them.[3] Tree graphs show the female–male difference across countries at each grade level.[4] At grade four and grade eight, separate trend graphs for each country show the percentage of students with misconceptions, misunderstandings, and errors over multiple assessment years.[5]

With the exception of the trend item results, all data reflect the most recent assessment in which each item was administered from 1995 to 2015, which for examples items, is the year the item was released (supplementary materials providing standard errors for all estimates are available for download at www.iea.nl/publications/RfEVol9).

4.2 Physics Results

We selected a set of 16 physics items from the TIMSS and TIMSS Advanced assessments from 1995 to 2015 that measure student understandings and misconceptions related to gravitational force. This item set includes four items from TIMSS Advanced, seven TIMSS items at grade eight, and six TIMSS items at grade four. We identified four performance objectives (POs) measured by these items, each with a specific set of related misconceptions and misunderstandings (Table 4.1).

We provide here a list of the full set of TIMSS and TIMSS Advanced items related to gravity (Table 4.2) organized by performance objective and grade level. This list shows the assessment year(s) when each item was administered, the item format (MC or CR), a brief item description, the figures where the items are shown in the report (released items only), and the specific type(s) of student misconceptions and

[2]Each country translates the international version of the TIMSS and TIMSS Advanced assessment items into their language(s) of instruction, and these translated national versions are verified by the IEA.

[3]As noted previously, results are based on both released and non-released items that measure the specific types of misconceptions, misunderstandings, and errors. Example items presented in the report are all restricted-use items (from 2015) and released items (from previous assessments). For trend items administered in multiple assessments, the data shown in the tables and figures in this section reflect the most recent assessment year (shown in Tables 4.2 and 4.22).

[4]Gender differences on TIMSS Advanced items are based on the sample of students who took the TIMSS Advanced assessment. As described in Chap. 3, the TIMSS Advanced population reflects a select group of students in each country, and the percentage of female and male students taking more advanced courses may differ from the percentages in the full population of students in their final year of secondary school (see Tables 3.4 and 3.5).

[5]Trend graphs are included only for grades four and eight, as there were no TIMSS Advanced trend items available that measured the specific physics and mathematics concepts under study.

Table 4.1 Physics performance objectives related to gravity with related misconceptions and misunderstandings, by grade level

Performance objective	Related misconceptions and/or misunderstandings	Grade level		
		TA	G8	G4
PO1: Determine the acceleration of thrown objects (after they are released)	Gravitational force (acceleration) acting on objects near Earth's surface is not constant but changes with the height of the object above the surface. (P1A)	✓		
	Objects thrown upward have no acceleration at their maximum height where the instantaneous velocity is zero (the instant it stops moving upward and reverses direction). (P1B)	✓		
	Gravitational acceleration is always in the direction of motion/velocity (rather than a constant acceleration directed toward the center of Earth). (P1C)	✓		
PO2: Determine the time duration between different points on the path of a thrown object	The time on the way up and the time on the way down are not equal (the downward acceleration due to gravity is not treated as constant). (P2)	✓		
PO3: Determine the effect of gravitational force on moving objects or on objects at rest	Gravity acts only on falling objects, but not on objects at rest (on the ground or sitting on another surface) or on objects that are moving upward. (P3A)	✓	✓	
	Gravity alone cannot cause an object initially at rest to start moving; it requires another force/push. (P3B)		✓	✓
PO4: Identify the direction of the force due to gravity	Gravitational force causes objects to fall "down" (in an "absolute downward" direction in space) rather than toward the center of Earth. (P4A)		✓	
	Gravity pushes upward on objects sitting on a solid surface and on objects that are moving upward. (P4B)			✓
	Gravity can move objects in other directions that are not "down" toward the surface of Earth. (P4C)			✓

Notes There are four physics performance objectives (PO1 to PO4). The related misconceptions and misunderstandings are coded (e.g., P1A, P1B, P2, etc.). The first two identifier codes refer to the corresponding physics objective number (e.g., P1, P2, etc.). When there is more than one misconception or misunderstanding related to the performance objective, a third identifier was added (i.e., A, B, C). Grade levels: TA = TIMSS Advanced, G8 = grade 8, G4 = grade 4

✓ Indicates that the misconception or misunderstanding was measured by one or more items at that grade level

Table 4.2 List of TIMSS and TIMSS Advanced physics items related to gravity, organized by performance objective and grade level

TIMSS item	Grade level	Assessment year(s)	Item format	Item description	Figure in this report	Physics misconceptions and misunderstandings								
						P1A	P1B	P1C	P2	P3A	P3B	P4A	P4B	P4C
Performance objective 1: Determine the acceleration of thrown objects (after they are released)														
Item 1A (PA33061A)	TIMSS Advanced	2015	MC	Motion of a ball thrown upward—acceleration at highest point	4.4	✓	✓							
Item 2 (PA13063)		1995	CR	Acceleration of a bouncing ball	4.5			✓						
Performance objective 2: Determine the time duration between different points on the path of a thrown object														
Item 1B (PA33061B)	TIMSS Advanced	2015	CR	Motion of a ball thrown upward—time between two points	4.4				✓					
Performance objective 3: Determine the effect of gravitational force on moving objects or on objects at rest														
Item 3 (PA23014)	TIMSS Advanced	2008	CR	Forces acting on a stone thrown upward	4.6					✓				
Item 4 (S032141)	Grade 8	2003, 2007, 2011	MC	Gravity acting on parachute jumper	4.7					✓				
Item 5 (S022012)		1999	MC	Gravity acting on a rocket being launched from Earth	–					✓				
Item 6 (S012075)		1995	MC	Gravity acting on an apple falling from a tree	–					✓				
Item 7 (S042211)		2007, 2011, 2015	CR	Forces acting on people sitting on a wall	4.8					✓				
Item 8 (S032281)		2003	MC	Why helium balloon moves upward	4.9					✓				
Item 9 (S042293A)		2007, 2011, 2015	CR	Force causing a ball thrown upward to fall	4.10						✓			

(continued)

Table 4.2 (continued)

TIMSS item	Grade level	Assessment year(s)	Item format	Item description	Figure in this report	Physics misconceptions and misunderstandings								
						P1A	P1B	P1C	P2	P3A	P3B	P4A	P4B	P4C
Performance objective 3: Determine the effect of gravitational force on moving objects or on objects at rest														
Item 10 (S031311)	Grade 4	2003, 2007, 2011	MC	Force causing an object to fall to the ground	4.12						✓			
Item 11 (S051147)		2011, 2015	CR	Force causing a marble to roll down a sloping track	4.13						✓			
Performance objective 4: Identify the direction of the force due to gravity														
Item 12 (S032714)	Grade 8	2003	MC	Direction gravity makes a ball fall at different places on Earth	4.11							✓		
Item 13 (S041119)	Grade 4	2007, 2011, 2015	MC	Direction of the force of Earth's gravity	4.14								✓	
Item 14 (S061048)[a]		2015	MC	Direction of movement due to gravity	–								✓	✓
Item 15 (S041308)		2007	MC	Example of an object moving due to gravity	4.15									✓
Item 16 (S031313)		2003	MC	Force that makes objects repel each other	4.16									✓

Notes See Table 4.1 for a full explanation of the codes related to physics misconceptions and misunderstandings. For each item, the relevant TIMSS item ID code as used in the TIMSS and TIMSS Advanced databases and released item sets is shown in brackets. Item 1A contributes to two misconceptions depending on the response options considered: misconception P1A includes students who selected either option B or C, while misconception P1B includes students who selected option A. Item 12 measures physics concepts related to gravity, but is classified as an earth science item in the TIMSS framework. Item 14 contributes to two misconceptions depending on the response options considered: misconception P4B includes students who selected option C, while misconception P4C includes students who selected either option B or D

✓ Indicates the specific type of physics misconception or misunderstanding is measured by the item; *MC* multiple choice; *CR* constructed response

[a]Item is secured

misunderstandings measured by each item. All physics results reported in this section are based on student performance on this set of items. (See Appendix Table A.1 for additional information on the physics items used in this study, including the specific response options or score categories used to determine the percentage of students demonstrating each type of misconception or misunderstanding.)

4.2.1 Student Performance on TIMSS and TIMSS Advanced Items Related to Gravity

The performance of students on the set of gravity items at each grade level covered a broad range both within and across countries (Figs. 4.1, 4.2, and 4.3), with some difficult items (<30% of students were correct) and some easier items (>70% of students were correct).[6] Average item performance across the five countries for TIMSS Advanced ranged from 25 to 54% correct, compared with a range of 36–79% correct at grade eight and 31–78% correct at grade four. Based on the average performance, the range of item percent correct was lower in TIMSS Advanced (25–42% correct) than at grade eight (36–79% correct) or grade four (31–78% correct). Across the set of gravity items at each grade level, at least half of students provided a correct response on five out of seven grade eight items and three out of six grade four items, compared to only one out of the four TIMSS Advanced items. However, there were notable differences in performance patterns observed across the five countries.

Across the four TIMSS Advanced items (Fig. 4.1), the broadest range of item performance was in Slovenia (from 16 to 72% correct) and the United States (from 15 to 64% correct). In contrast, item-level performance ranged from 23 to 49% correct in the Russian Federation and, in Norway, was clustered between 46 and 63% correct. Performance was lowest in Italy, with the item percent correct ranging from 9 to 31%, and generally higher in Norway when considering the full set of items. The most difficult item in all countries was item 2 ("acceleration of a bouncing ball"), a CR item under performance objective 1 ("determine the acceleration due to gravity of thrown objects (after they are thrown)").[7] In comparison, item 1B ("motion of a ball thrown upward–time between two points"), a CR item from performance objective 2 ("determine the time duration between different points on the path of a thrown object"), was the easiest item for all countries except Italy, with at least 60% of students correct in Norway (63%), the United States (64%), and Slovenia (72%).

[6]The data (Figs. 4.1, 4.2, and 4.3) reflect the most recent assessment in which each item was administered from 1995 to 2015 (see Table 4.2 for the most recent assessment for each item). Changes in performance between assessment cycles for trend items are reported later (Figs. 4.23 and 4.24).

[7]Item 2 is a released item from the 1995 assessment. Comparable data are not available from Italy. Thus, this item is only included for the other four countries (Norway, the Russian Federation, Slovenia, and the United States).

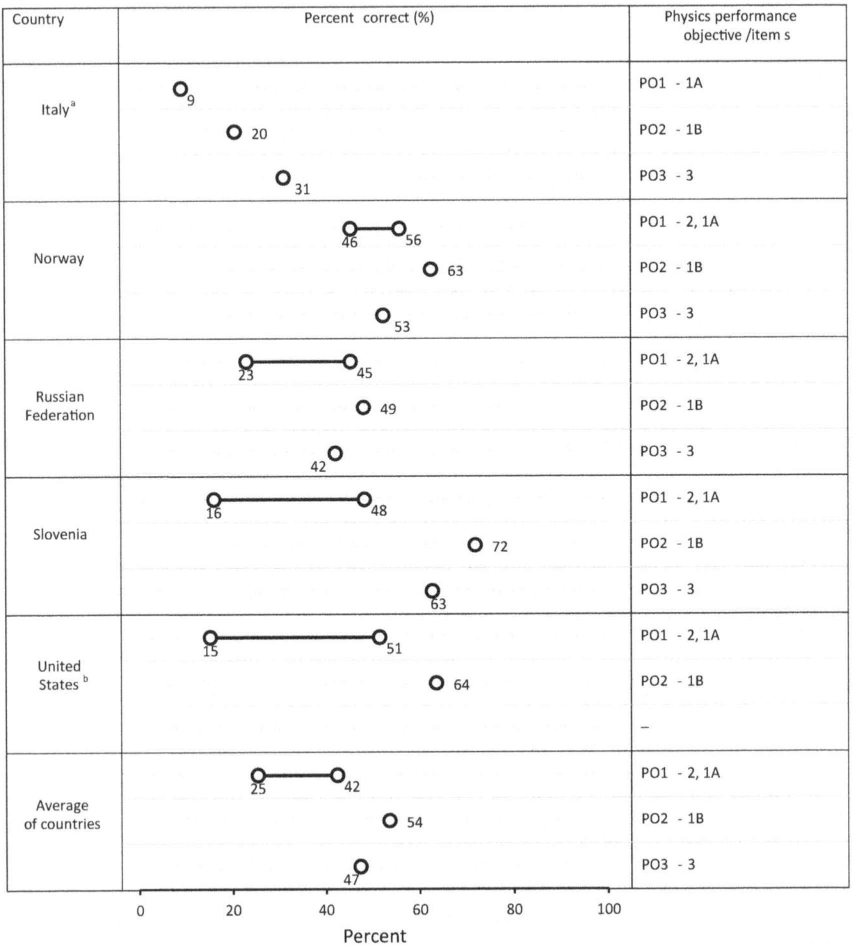

Fig. 4.1 Student performance on TIMSS Advanced physics items, by country and performance objective, 1995, 2008, and 2015. *Notes* Percent correct is the percentage of students receiving credit on each item. For MC and short CR items (each worth one score point), this reflects the percentage of students who provided a correct answer. For extended CR items, this reflects the weighted percentage of students receiving full credit (2 points) or partial credit (1 point). The percentages are for the most recent cycle each item was administered. Data for items 1A and 1B are from 2015; data for item 3 are from 2008, and data for item 2 are from 1995. Physics performance objectives (PO): PO1 = determine the acceleration of thrown objects (after they are released), PO2 = determine the time duration between different points on the path of a thrown object, PO3 = determine the effect of gravitational force on moving objects or on objects at rest. [a]Data not available for item 2 (see Appendix for country-specific notes). [b]Data not available for item 3 (see Appendix for country-specific notes)

At grade eight (Fig. 4.2), a broad range of item performance was found in all five countries and especially in Norway (from 32 to 85% correct). Three MC items, namely items 4, 5, and 6 from performance objective 3 ("determine the effect of gravitational force on moving objects or on objects at rest"), were among the most difficult items in all countries. In particular, performance on item 4 ("gravity acting on a parachute jumper") ranged from 26% correct in Italy to 30–40% correct in Norway, the Russian Federation, and the United States, to 47% correct in Slovenia. In comparison, the easiest item was item 12 ("direction gravity makes a ball fall at different places on Earth") from performance objective 4 ("identify the direction of the force due to gravity"), with at least 70% correct in all five countries, and more than 80% correct in Norway, the Russian Federation, and Slovenia.

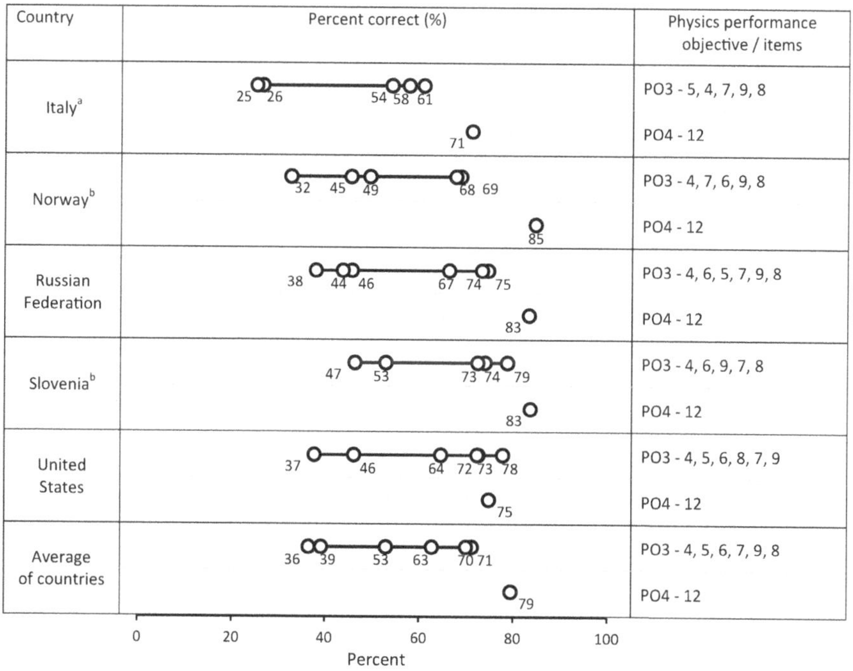

Fig. 4.2 Student performance on TIMSS grade eight physics items, by country and performance objective, 1995, 1999, 2003, 2011, and 2015. *Notes* Percent correct is the percentage of students receiving credit on each item. For MC and short CR items (each worth one score point), this reflects the percentage of students who provided a correct answer. For extended CR items, this reflects the weighted percentage of students receiving full credit (2 points) or partial credit (1 point). The percentages are for the most recent cycle each item was administered. Data for items 7 and 9 are from 2015; data for item 4 are from 2011; data for items 8 and 12 are from 2003; data for item 5 are from 1999; and data for item 6 are from 1995. Physics performance objectives (PO): PO3 = determine the effect of gravitational force on moving objects or on objects at rest, PO4 = identify the direction of the force due to gravity. [a]Data not available for item 6 (see Appendix for country-specific notes). [b]Data not available for item 5 (see Appendix for country-specific notes)

The broadest range of item performance was found at grade four (Fig. 4.3) in Italy, the Russian Federation, and Slovenia, compared to Norway (where the range was greatest at grade eight) and the United States (where the range was more similar across the three grade levels). In particular, item performance in Slovenia ranged from 16 to 88% correct (a spread of 72 percentage points at grade four). Item 11 from performance objective 3 ("determine the effect of gravitational force on moving objects or on objects at rest") and items 14 and 16 from performance objective 4 ("identify the direction of the force due to gravity"), were among the most difficult items in all countries, although the specific pattern of performance varied. The CR item 11 ("force causing a marble to roll down a sloping track") ranged from <20% correct in Italy (15%) and Slovenia (16%), to 29% correct in Norway, to ≥40% correct in the Russian Federation (53%) and the United States

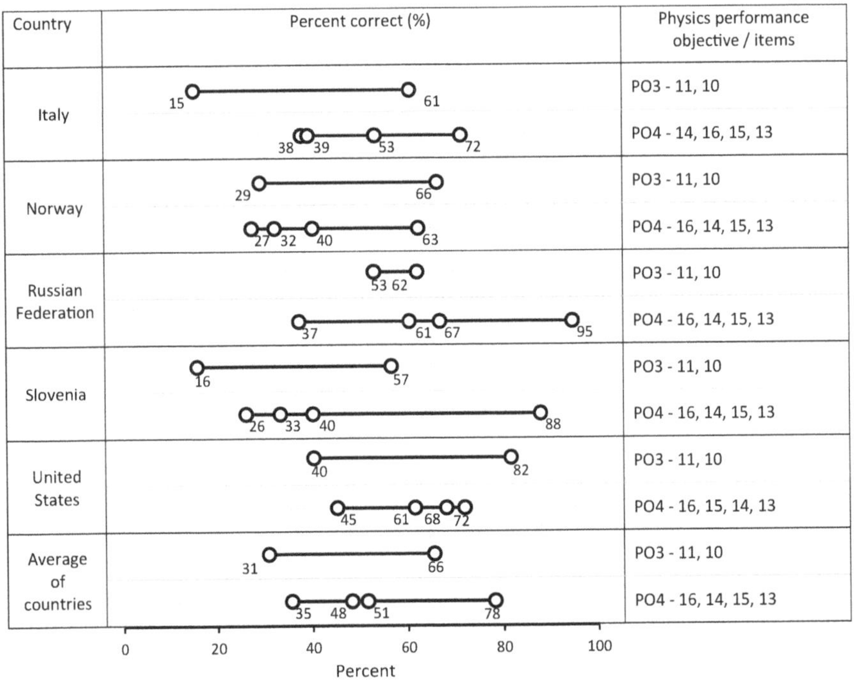

Fig. 4.3 Student performance on TIMSS grade four physics items, by country and performance objective, 2003, 2007, 2011, and 2015. *Notes* Percent correct is the percentage of students receiving credit on each item. For MC and short CR items (each worth one score point), this reflects the percentage of students who provided a correct answer. For extended CR items, this reflects the weighted percentage of students receiving full credit (2 points) or partial credit (1 point). The percentages are for the most recent cycle each item was administered. Data for items 11, 13, and 14 are from 2015; data for item 10 are from 2011; data for item 15 are from 2007; and data for item 16 are from 2003. Physics performance objectives (PO): PO3 = determine the effect of gravitational force on moving objects or on objects at rest, PO4 = identify the direction of the force due to gravity

(40%). Performance on the MC item 14 ("direction of movement due to gravity") ranged from 32% correct in Norway to 68% correct in the United States, and performance on MC item 16 ("force that makes objects repel each other") ranged from 26% correct in Slovenia to 45% correct in the United States. In comparison, item 10 ("force causing an object to fall to the ground") from performance objective 3 and item 13 ("direction of the force of Earth's gravity") from performance objective 4 (both MC items) were amongst the easiest items across all countries, with $\geq$60% of students responding correctly to both items (except in Slovenia, with 57% correct on item 10). Performance on item 10 was particularly high in the United States (82% correct) and performance on item 13 was particularly high in the Russian Federation (95% correct) and Slovenia (88% correct).

4.2.2 *Common Types of Misconceptions and Misunderstandings Related to Gravity Across Countries*

A key understanding is that the force due to Earth's gravity acting on an object on or near Earth's surface is constant, resulting in a constant acceleration (approximately 10 m s^{-2}) directed toward the center of Earth. By the end of secondary school, students are expected to understand that the only forces acting on a thrown object (after it is released) are the downward force due to gravity and air resistance and that the observed motion (slowing, reaching a maximum height, and then falling back down) is the result of the constant acceleration due to gravity at all positions in the path of the object. The first example item (item 1: Fig. 4.4, and Tables 4.3 and 4.4) requires students to apply Newton's laws of motion to answer two questions about the motion of a ball thrown vertically upward, and shows that many TIMSS Advanced students had difficulty applying these concepts.

Part A is a MC item requiring students to identify the acceleration of the ball at its highest position (the instant it stops moving upward and reverses direction). A correct response to part A (option D) requires students to know that the acceleration due to gravity is constant and applies equally to the ball at all positions. Across the five countries included in the study, the percent correct ranged from 9% (Italy) to 56% (Norway), with an international average of 42% correct. On average, nearly half of students internationally (48%) indicated that the acceleration was zero (option A), demonstrating the misconception that there is no acceleration since the instantaneous velocity at that position is zero (rather than a constant acceleration due to gravity at all positions). This misconception was less common in Norway (39%), but this still reflects more than one-third of students.

Another 8% of students on average across countries incorrectly determined that the magnitude of the acceleration due to gravity was different at point 3 (either half or twice that at point 2), demonstrating the misconception that the force of gravity changed with the height of the ball (options B and C). Students selecting these options may be confusing gravitational force with gravitational potential energy

Sally throws a ball vertically upward as shown. The ball moves from her hand at point 1 to a maximum height at point 3. Point 2 is halfway between points 1 and 3. The ball has an acceleration of −10 m/s² at point 2.

A. What is the acceleration at point 3 at the instant between its upward motion and downward motion? Disregard air resistance.

Ⓐ zero m/s²

Ⓑ $\frac{-10}{2}$ m/s²

Ⓒ 2(−10) m/s²

Ⓓ −10 m/s²

B. How does the time duration between points 2 and 3 on the way up compare to the time duration between points 3 and 2 on the way down? Disregard air resistance.

Item information

Item ID:
PA33061 (A and B)

Year(s) administered:
2015

Performance objective:
Part A. Determine the acceleration of thrown objects (after they are released)

Part B. Determine the time duration between different points on the path of a thrown object

Correct answer:
Part A. D

Scoring guide for Part B

Correct response	
10	Indicated that times are equal
Incorrect response	
79	Incorrect (including crossed out, erased, stray marks, illegible, or off task)
Non response	
99	Blank

Fig. 4.4 TIMSS Advanced physics item 1, 2015. *Source* TIMSS Advanced 2015 Assessment. Copyright © 2017 International Association for the Evaluation of Educational Achievement (IEA). TIMSS & PIRLS International Study Center, Lynch School of Education, Boston College

($PE_g = mgh$), which increases with height (h), or incorrectly applying Newton's law of universal gravitation ($F_g = Gm_1m_2/r^2$), where gravitational force (F_g) decreases with the distance squared (r^2). Although this law can be applied to objects that are far from Earth, the difference in force or acceleration due to a change in height is negligible for objects near Earth's surface, where the acceleration due to gravity (g) is treated as a constant.

In part B, students were asked to determine the time duration between two points on the path of the ball (halfway up and halfway down). A correct response to part B requires students to indicate that the time traveled by the ball is the same on the way up as it is on the way down, as shown in the scoring guide. This relationship can be determined by applying Newton's laws of motion to the situation where there is a constant acceleration due to gravity (g), and TIMSS Advanced students are expected to have covered this in their physics courses. A common misconception, though, is that the time on the way down is shorter because the ball is accelerating (speeding up) on the way down and decelerating (slowing down) on the way up (i.e., the downward acceleration due to gravity is not treated as constant). On average, slightly more than half (54%) of TIMSS Advanced students answered part B correctly. Most students in Norway (63%), Slovenia (72%), and the United States (64%) provided a correct response. In comparison, about half of the TIMSS Advanced students in the Russian Federation (49%) and about one-fifth in Italy (20%) did so.

Table 4.3 Student performance data for physics item 1A (PA33061A), 2015

Country	Percentage of students (%)					
	Correct (D)	A	B	C	D	Omitted
Italy	9 ↓	71	8	8	9	4
Norway	56 ↑	39	2	2	56	0
Russian Federation	45	48	3	4	45	1
Slovenia	48 ↑	44	3	4	48	1
United States	51 ↑	41	3	4	51	1
Average of countries	42	48	4	4	42	2

Notes Detail may not sum to totals due to rounding
↑ Significantly higher than the average of countries
↓ Significantly lower than the average of countries

Table 4.4 Student performance data for physics item 1B (PA33061B), 2015

Country	Percentage of students (%)			
	Correct (10)	10	79	99
Italy	20 ↓	20	28	51
Norway	63 ↑	63	22	16
Russian Federation	49 ↓	49	25	26
Slovenia	72 ↑	72	21	6
United States	64 ↑	64	34	2
Average of countries	54	54	26	20

Notes Detail may not sum to totals due to rounding
↑ Significantly higher than the average of countries
↓ Significantly lower than the average of countries

In the second TIMSS Advanced example item (item 2: Fig. 4.5 and Table 4.5), students were asked to draw arrows on the figure of a bouncing ball that represent the direction of acceleration of the ball at three positions above the floor (where the only force acting on the ball is due to gravity). Approximately one-quarter of TIMSS Advanced students (25% on average) correctly indicated that the acceleration is directed downward in all three positions (code 10 in the scoring guide).[8] On

[8]Item 2 is a released TIMSS Advanced item from 1995. Data for Italy are not available; thus, the international average is based on the other four countries (Norway, the Russian Federation, Slovenia, and the United States).

The figure shows the trajectory of a ball bouncing on a floor, with negligible air resistance.

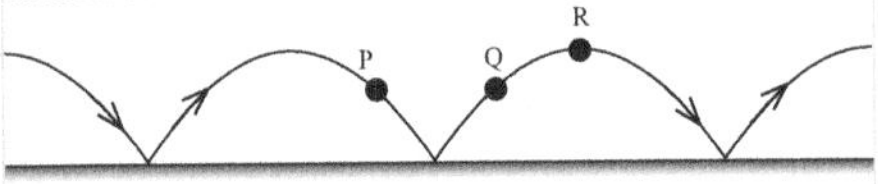

Draw arrows on the figure showing the direction of the acceleration of the ball at points P, Q and R.

Item information

Item ID:
PA13063

Year(s) administered: 1995

Performance objective:
Determine the acceleration of thrown objects (after they are released)

Scoring guide

Correct response	
10	The acceleration is parallel to g, downwards arrows at P, Q, and R. (See following diagrams)
Incorrect response	
70	The acceleration is parallel to g, downwards arrow at P, upwards at Q, and zero at R
71	The acceleration is parallel to g, downwards arrow at P, upwards at Q, either upwards or downwards at R
72	The acceleration has the same direction as the motion (at least P and Q). Any response at R
73	The acceleration has the same direction as the motion at P, the opposite direction from the motion at Q. Any response at R
74	The acceleration has the direction perpendicular to the motion (at least at P and Q)
79	Other incorrect responses
Non response	
90	Crossed-out/erased, illegible, or impossible to interpret
99	Blank

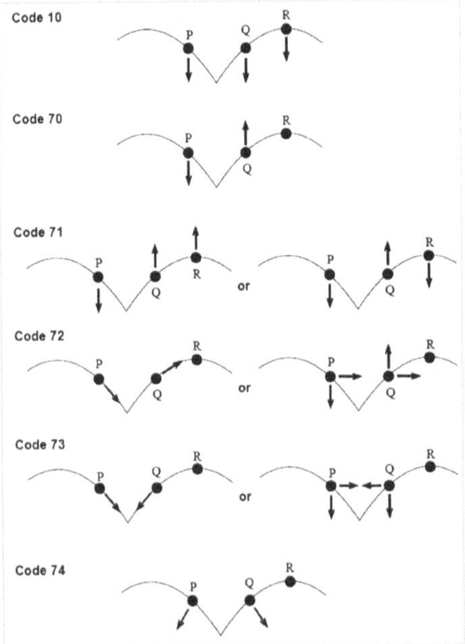

Fig. 4.5 TIMSS Advanced physics item 2, 1995. *Source* TIMSS Advanced 1995 Assessment. Copyright © 1997 International Association for the Evaluation of Educational Achievement (IEA). TIMSS & PIRLS International Study Center, Lynch School of Education, Boston College

Table 4.5 Student performance data for physics item 2 (PA13063), 1995

Country	Percentage of students (%)									
	Correct (10)		10	70	71	72	73	74	79	99
Italy	—		—	—	—	—	—	—	—	—
Norway	46	⋂	46	6	7	14	4	7	14	2
Russian Federation	23		23	3	2	29	6	17	15	5
Slovenia	16	⋃	16	13	2	29	9	8	17	7
United States	15	⋃	15	13	2	38	4	1	27	0
Average of countries	25		25	9	3	27	6	8	18	4

Notes Detail may not sum to totals due to rounding
⋂ Significantly higher than the average of countries
⋃ Significantly lower than the average of countries
— Data not available for item 2 (see Appendix for country-specific notes)

average, about one-quarter of students (27%) demonstrated the common misconception that the acceleration is in the same direction as the motion of the ball (along the curved path at points P and Q) and that there is no acceleration when the ball is at its maximum height (point R) (code 72 in the scoring guide). This misconception was most common in the United States (38%) and least common in Norway (14%). Another 12% internationally indicated that the acceleration of the ball is upward on its way up (point Q) and downward on its way down (point P), with either an upward or downward acceleration or no acceleration at point R (codes 70 and 71). In addition, about 8% internationally indicated that the acceleration is perpendicular to the direction of motion at points P and Q (code 74). This type of response reflects a misunderstanding that the acceleration of the ball moving along a curved path is caused by a centripetal force directed toward its center (like objects orbiting the Earth). The frequency of this misconception ranged from 1% of students in the United States to 17% of students in the Russian Federation. Another 24% of students on average provided other types of incorrect responses (codes 73 and 79), and about 4% left the item blank.

In the final TIMSS Advanced item (item 3: Fig. 4.6 and Table 4.6), about three-quarters of students (75% on average across countries) identified gravity as a force acting on a stone after it was thrown straight up in the air (codes 10 and 70 in the scoring guide).[9] The percentage of students who did not identify gravity ranged from almost 3% in Slovenia[10] to 42% in the Russian Federation. These percentages include incorrect responses (codes 71 and 79), as well as students who left the item blank.

[9]Item 3 is a released TIMSS Advanced item from 2008. The United States did not participate in the 2008 assessment; thus, the international average is based on the other four countries (Italy, Norway, the Russian Federation, and Slovenia).

[10]This is based on the unrounded data (not shown in Table 4.6). Data to the nearest 0.01% are available at www.iea.nl/publications/RfEVol9.

Lisa threw a small stone straight up into the air. What forces act on the stone after it was thrown?

Item information
Item ID:
PA23014

Year(s) administered:
2008

Performance objective:
Determine the effect of gravitational force on moving objects or on objects at rest

Scoring guide

Correct response	
10	Gravity/weight and air resistance
Incorrect response	
70	Gravity/weight mentioned, but not air resistance
71	Air resistance mentioned, but not gravity/weight
79	Other incorrect (including crossed out, erased, stray marks, illegible, or off task)
Non response	
90	Crossed-out/erased, illegible, or impossible to interpret
99	Blank

Fig. 4.6 TIMSS Advanced physics item 3, 2008. *Source* TIMSS 2008 Assessment. Copyright © 2010 International Association for the Evaluation of Educational Achievement (IEA). TIMSS & PIRLS International Study Center, Lynch School of Education, Boston College

Table 4.6 Student performance data for physics item 3 (PA23014), 2008

Country	Percentage of students (%)					
	Correct (10)	10	70	71	79	99
Italy	31 ⇓	31	36	0	20	12
Norway	53 ⇑	53	25	0	21	1
Russian Federation	42 ⇓	42	16	1	34	7
Slovenia	63 ⇑	63	34	0	2	0
United States	—	—	—	—	—	—
Average of countries	47	47	28	0	19	5

Notes Detail may not sum to totals due to rounding
⇑ Significantly higher than the average of countries
⇓ Significantly lower than the average of countries
— Data not available for item 3 (see Appendix for country-specific notes)

By grade eight, students are expected to be able to determine the effect of gravitational force acting on moving objects or on objects at rest. However, many grade eight students demonstrated the misconception that the force of gravity acts only on falling objects, not on objects that are at rest. This is shown in the first grade eight TIMSS example item (item 4: Fig. 4.7 and Table 4.7), which asks students to identify at which position(s) the force of gravity acts on a parachute jumper (in the

The figure shows a parachute jumper in four positions.

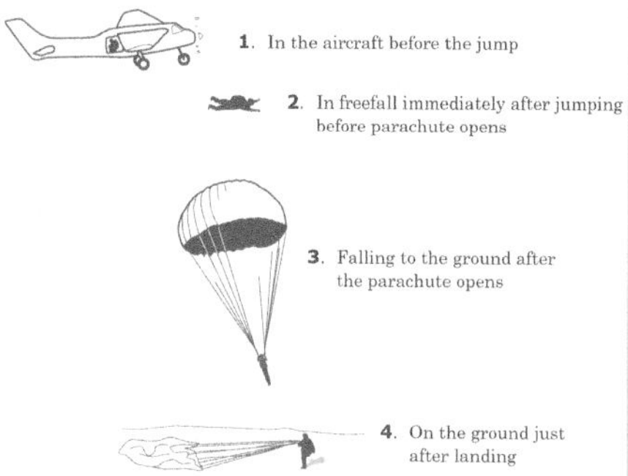

In which of the positions does the force of gravity act on the jumper?

Ⓐ Position 2 only.

Ⓑ Positions 2 and 3 only.

Ⓒ Positions 1, 2 and 3 only.

Ⓓ Positions 1, 2, 3, and 4.

Item information

Item ID:
S032141

Year(s) administered: 2011, 2007, and 2003

Performance objective:
Determine the effect of gravitational force on moving objects or on objects at rest

Correct answer:
D

Fig. 4.7 TIMSS grade eight physics item 4, 2011. *Source* TIMSS 2011 Assessment. Copyright © 2013 International Association for the Evaluation of Educational Achievement (IEA). TIMSS & PIRLS International Study Center, Lynch School of Education, Boston College

Table 4.7 Student performance data for physics item 4 (S032141), 2011

Country	Percentage of students (%)					
	Correct (D)	A	B	C	D	Omitted
Italy	26 ↓	17	51	5	26	1
Norway	32	11	47	7	32	2
Russian Federation	38	16	41	4	38	1
Slovenia	47 ↑	8	38	7	47	0
United States	37	8	48	6	37	1
Average of countries	36	12	45	6	36	1

Notes Detail may not sum to totals due to rounding
↑ Significantly higher than the average of countries
↓ Significantly lower than the average of countries

aircraft prior to jumping, in freefall, falling with an open parachute, and on the ground after landing). On average across countries, 36% of students correctly identified that the force of gravity was acting on the jumper at all four positions (option D), while over half (57%) indicated that gravity acted only when the jumper was falling (specifically, 45% falling with the parachute open or closed (option B) and 12% in freefall only (option A)). This misconception was common across all five countries, although somewhat less frequent in Slovenia (46%) and more frequent in Italy (68%). The same type of common misconception was demonstrated across countries on two other TIMSS grade eight items involving the application of the same concept in different contexts[11]: a rocket being launched from Earth (item 5); and an apple falling from a tree (item 6). In the context of a rocket launch (item 5), about one-third of grade eight students on average (36%)[12] indicated that gravity acts on the rocket only when it is falling back to Earth and not when it is sitting on the launch pad or moving upward after being launched. In the context of an apple falling from a tree (item 6), 40% of students on average[13] indicated that gravity acts on the apple only while it is falling or still hanging from the tree, and not once it lands on the ground.

The second grade eight example (item 7: Fig. 4.8 and Table 4.8) also showed that many students had a lack of understanding about gravity acting on objects at rest. The item asked students to identify the forces acting on two people sitting on a wall. A complete response (code 10 in the scoring guide) must identify two balanced forces: the downward force due to gravity and the upward force from the wall. On average across the five countries, 63% of students received credit for providing a response that includes gravity and/or the upward force from the wall (codes 10, 11, 12, and 19 in the scoring guide), but only 9% referred correctly to both forces (codes 10 and 11). A complete response was most common in Slovenia (26% of students). On average across countries, about half (51%) of students included only one force, with most of these responses referring to gravity alone. The percentage of students who did not identify gravity (codes 70, 71, 79, and 99) ranged from about a quarter of students in Slovenia (27%) and the United States (26%), to more than half of students in Norway (55%). Some students indicated that there were no forces acting on the people, and others referred to gravity pushing up. These findings at grade eight are similar to results from a previous study of the misconceptions of students in elementary school (Darling 2012). In that study, many students thought that there were no forces acting on a book sitting at rest on a desk.

In the next grade eight example (item 8: Fig. 4.9 and Table 4.9), students were asked to identify the best explanation for why a helium balloon moves upward when it is released. While most students (71% on average across countries) responded correctly that the density of helium is less than the density of air (option A), the most common incorrect response in all countries except Slovenia was that there is no gravity acting on helium balloons (option C), chosen by 14% of students on average across countries.

---

[11]These two items are not shown as exhibits in the report, but both are released items available from the IEA website (see www.iea.nl).

[12]Item 5 is a released TIMSS item from 1999. Norway did not participate in the 1999 assessment, and comparable data are not available for Slovenia. Thus, the international average is based on the other three countries (Italy, the Russian Federation, and the United States).

[13]Item 6 is a released TIMSS item from the 1995 assessment. Comparable data are not available for Italy. Thus, the international average is based on the other four countries (Norway, the Russian Federation, Slovenia, and the United States).

Bernie and Travis are sitting on a wall.

Are any forces acting on them?

(Check one box.)

☐ Yes

☐ No

Explain your answer.

Item information

Item ID:
S042211

Year(s) administered:
2015, 2011, 2007

Performance objective:
Determine the effect of gravitational force on moving objects or on objects at rest

Scoring guide

Correct response	
10	**Yes** and refers to the following **two** forces on the children: gravity (down) **AND** the wall (up)
11	**Yes** and refers to these **two** forces on the wall: (children's) weight (down) **AND** the wall/ground (up)
12	**Yes** and refers to **one** force: gravity (down) **OR** wall/ground (up) **OR** weight (down)
19	Other correct
Incorrect response	
70	**Yes** with an explanation that only includes friction
71	**No** with or without explanation
79	Other incorrect (including crossed out, erased, stray marks, illegible, or off task)
Non response	
99	Blank

Fig. 4.8 TIMSS grade eight physics item 7, 2015. *Source* TIMSS 2015 Assessment. Copyright © 2017 International Association for the Evaluation of Educational Achievement (IEA). TIMSS & PIRLS International Study Center, Lynch School of Education, Boston College

Table 4.8 Student performance data for physics item 7 (S042211), 2015

Country	Percentage of students (%)								
	Correct (10-19)	10	11	12	19	70	71	79	99
Italy	54 ↓	2	0	48	4	1	15	25	5
Norway	45 ↓	0	0	40	5	0	12	34	9
Russian Federation	67	7	0	57	2	1	9	21	2
Slovenia	74 ↑	26	3	45	0	0	0	25	0
United States	73 ↑	6	1	64	2	1	15	11	1
Average of countries	63	8	1	51	2	0	10	23	3

Notes Detail may not sum to totals due to rounding. Although the table displays rounded data, the calculations of the combined correct responses are based on unrounded data

↑ Significantly higher than the average of countries

↓ Significantly lower than the average of countries

A balloon filled with helium gas is set free and starts to move upward. Which of the following best explains why the helium balloon moves upward?

Ⓐ The density of helium is less than the density of air.

Ⓑ The air resistance lifts the balloon up.

Ⓒ There is no gravity acting on helium balloons.

Ⓓ The wind blows the balloon upward.

Item information

Item ID:
S032281

Year(s) administered:
2003

Performance objective:
Determine the effect of gravitational force on moving objects or on objects at rest

Correct answer:
A

Fig. 4.9 TIMSS grade eight physics item 8, 2003. *Source* TIMSS 2003 Assessment. Copyright © 2005 International Association for the Evaluation of Educational Achievement (IEA). TIMSS & PIRLS International Study Center, Lynch School of Education, Boston College

Table 4.9 Student performance data for physics item 8 (S032281), 2003

Country	Percentage of students (%)							
	Correct (A)	A	B	C	D	Invalid	Omitted
Italy	61 ↓	61	12	19	4	1	3
Norway	69	69	10	16	3	1	1
Russian Federation	75 ↑	75	7	14	2	1	1
Slovenia	79 ↑	79	11	6	3	0	0
United States	72	72	12	13	2	0	1
Average of countries	71	71	10	14	3	0	1

Notes Detail may not sum to totals due to rounding
↑ Significantly higher than the average of countries
↓ Significantly lower than the average of countries

Another grade eight item (item 9: Fig. 4.10 and Table 4.10) asked students to identify the force that causes a ball thrown upward to fall from its highest point. Gravity causing objects to fall is an expectation even at the grade four level, and most grade eight students (70% on average) correctly identified gravity (code 10 in the scoring guide). However, some grade eight students may have difficulty with the concept that it is gravity alone acting on the ball after it is thrown upward that makes it reverse direction and start falling back down. On average, 30% of grade eight students provided an incorrect response or left the item blank.

In the final grade eight example (item 12: Fig. 4.11 and Table 4.11), students were asked to identify the direction gravity makes a ball fall at three different places on Earth. While 79% of students on average across countries correctly determined that the ball would fall toward the surface of Earth at all three locations (option D), some students (14% on average) determined that the ball would always fall "down" relative to the bottom of the page rather than toward Earth's surface (option A).

Jeffrey throws a ball up into the air, as shown in the diagram. It reaches its highest point at X and then falls straight down to the ground at point Y. The ball then bounces straight up again.

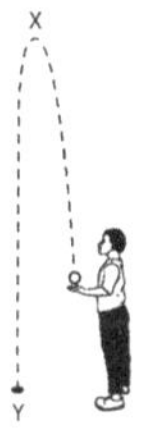

A. What force causes the ball to fall from point X to point Y?

Item information

Item ID:
S042293A

Year(s) administered:
2015, 2011, 2007

Performance objective:
Determine the effect of gravitational force on moving objects or on objects at rest

Scoring guide

Correct response	
10	Gravity (gravitational pull, gravitational force)
Incorrect response	
79	Incorrect (including crossed out, erased, stray marks, illegible, or off task)
Non response	
99	Blank

Fig. 4.10 TIMSS grade eight physics item 9, 2015. *Source* TIMSS 2015 Assessment. Copyright © 2017 International Association for the Evaluation of Educational Achievement (IEA). TIMSS & PIRLS International Study Center, Lynch School of Education, Boston College

Table 4.10 Student performance data for physics item 9 (S042293A), 2015

Country	Percentage of students (%)				
	Correct (10)		10	79	99
Italy	58	⬇	58	15	28
Norway	68		68	14	18
Russian Federation	74		74	6	20
Slovenia	73		73	22	6
United States	78	⬆	78	21	1
Average of countries	70		70	15	15

Notes Detail may not sum to totals due to rounding
⬆ Significantly higher than the average of countries
⬇ Significantly lower than the average of countries

Another 6% of students selected option B or C, both indicating that the ball falls "down" from the position at the "bottom" of the Earth. This indicates a lack of understanding demonstrated by some grade eight students about the direction of gravitational force that is expected by grade four (that gravity pulls objects toward Earth). These results are similar to those from an earlier international study (Sneider and Pulos 1983), which reported that about one-fifth of 13 to 14-year-old students (from middle schools in California, USA, and Jerusalem, Israel) demonstrated the

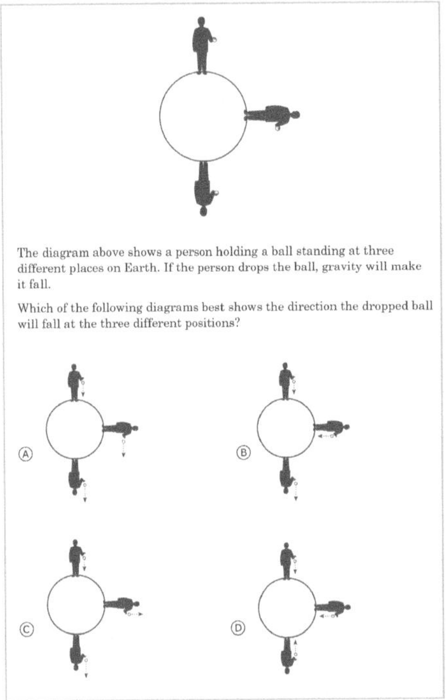

Item information
Item ID:
S032714

Year(s) administered:
2003

Performance objective:
Identify the direction of the force due to gravity

Correct answer:
D

Fig. 4.11 TIMSS grade eight physics item 12, 2003. *Source* TIMSS 2003 Assessment. Copyright © 2005 International Association for the Evaluation of Educational Achievement (IEA). TIMSS & PIRLS International Study Center, Lynch School of Education, Boston College

Table 4.11 Student performance data for physics item 12 (S032714), 2003

Country	Percentage of students (%)						
	Correct (D)	A	B	C	D	Invalid	Omitted
Italy	71 ▼	19	5	3	71	0	2
Norway	85 ▲	8	2	4	85	1	0
Russian Federation	83 ▲	13	1	2	83	1	1
Slovenia	83 ▲	11	2	1	83	0	2
United States	75 ▼	17	4	3	75	0	0
Average of countries	79	14	3	3	79	0	1

Notes Detail may not sum to totals due to rounding
▲ Significantly higher than the average of countries
▼ Significantly lower than the average of countries

misconception that objects fall toward an "absolute down" in space. According to the Sneider and Pulos study, while most students knew that objects fall toward the ground (surface of Earth), only about half understood that this is because the force of gravity is directed toward the center of Earth.

At grade four, students are assessed on their knowledge that gravity is the force that draws objects to Earth. In the first TIMSS grade four example item (item 10: Fig. 4.12 and Table 4.12), about two-thirds of students (66% on average) correctly chose gravity as the force that causes an object to fall (from among a list of given forces). However, some students (15% on average across countries) indicated that it is a push from the hand that causes the object to fall (option D). The percentage of students demonstrating this misconception ranged from 7% in the United States, to 24% in Italy.

What causes an object to fall to the ground when you let it drop from your hand?

Ⓐ magnetism
Ⓑ gravity
Ⓒ air resistance
Ⓓ the push from your hand

Item information
Item ID:
S031311

Year(s) administered:
2011, 2007, 2003

Performance objective:
Determine the effect of gravitational force on moving objects or on objects at rest

Correct answer:
B

Fig. 4.12 TIMSS grade four physics item 10, 2011. *Source* TIMSS 2011 Assessment. Copyright © 2013 International Association for the Evaluation of Educational Achievement (IEA). TIMSS & PIRLS International Study Center, Lynch School of Education, Boston College

Table 4.12 Student performance data for physics item 10 (S031311), 2011

Country	Percentage of students (%)						
	Correct (B)		A	B	C	D	Omitted
Italy	61	↓	7	61	5	24	3
Norway	66		11	66	7	12	4
Russian Federation	62		7	62	10	19	2
Slovenia	57	↓	17	57	10	14	2
United States	82	↑	3	82	6	7	2
Average of countries	66		9	66	8	15	3

Notes Detail may not sum to totals due to rounding
↑ Significantly higher than the average of countries
↓ Significantly lower than the average of countries

In the second grade four example (item 11: Fig. 4.13 and Table 4.13), students were shown an image of a marble placed at the top of a sloping track and asked to name the force that causes the marble to roll down the track. Less than one-third of students on average across countries (31%) correctly named gravity as the force that moves the marble (code 10 in the scoring guide), with performance lowest in Italy (15% correct) and Slovenia (16% correct) and highest in the Russian Federation (53% correct). About half of students (53% on average) provided an incorrect

Marcus puts a marble at the top of a sloping track as shown below.

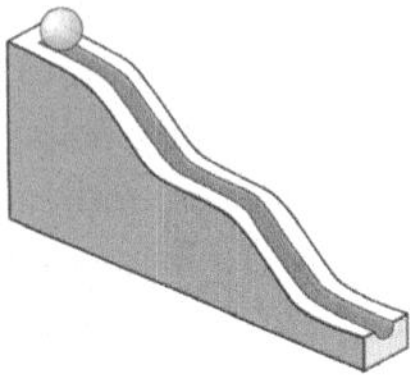

The marble rolls down the track.

Name the force that moves the marble.

Item information
Item ID:
S051147

Year(s) administered:
2015, 2011

Performance objective:
Determine the effect of gravitational force on moving objects or on objects at rest

Scoring guide

Correct response	
10	Gravity (explicitly or implicitly)
Incorrect response	
79	Incorrect (including crossed out, erased, stray marks, illegible, or off task)
Non response	
99	Blank

Fig. 4.13 TIMSS grade four physics item 11, 2015. *Source* TIMSS 2015 Assessment. Copyright © 2017 International Association for the Evaluation of Educational Achievement (IEA). TIMSS & PIRLS International Study Center, Lynch School of Education, Boston College

Table 4.13 Student performance data for physics item 11 (S051147), 2015

Country	Percentage of students (%)			
	Correct (10)	10	79	99
Italy	15 ↓	15	56	29
Norway	29	29	51	20
Russian Federation	53 ↑	53	38	9
Slovenia	16 ↓	16	65	19
United States	40 ↑	40	55	5
Average of countries	31	31	53	16

Notes Detail may not sum to totals due to rounding
↑ Significantly higher than the average of countries
↓ Significantly lower than the average of countries

response (code 79), with many students attributing the motion of the marble to another force, such as the wind or a push from the hand, and another 16% left the item blank.

Performance on the next grade four example (item 13: Fig. 4.14 and Table 4.14) showed that while most students (78% on average across countries) correctly indicated that the force of Earth's gravity acting on a block sitting on a table is in the downward direction (option C), some students (13% on average) demonstrated the misconception that the direction of the force of Earth's gravity is upward (away from the table top; option A). This misconception was most frequent in Norway

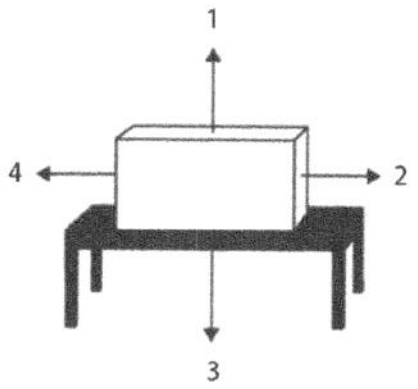

Look at the block on the table.

Which arrow shows the direction of the force of Earth's gravity?

Ⓐ 1
Ⓑ 2
Ⓒ 3
Ⓓ 4

Item information
Item ID:
S041119

Year(s) administered:
2015, 2011, 2007

Performance objective:
Identify the direction of the force due to gravity

Correct answer:
C

Fig. 4.14 TIMSS grade four physics item 13, 2015. *Source* TIMSS 2015 Assessment. Copyright © 2017 International Association for the Evaluation of Educational Achievement (IEA). TIMSS & PIRLS International Study Center, Lynch School of Education, Boston College

Table 4.14 Student performance data for physics item 13 (S041119), 2015

Country	Percentage of students (%)						
	Correct (C)		A	B	C	D	Omitted
Italy	72	▼	18	4	72	3	3
Norway	63	▼	24	6	63	3	4
Russian Federation	95	▲	3	1	95	1	0
Slovenia	88	▲	4	1	88	4	3
United States	72	▼	18	3	72	5	3
Average of countries	78		13	3	78	3	3

Notes Detail may not sum to totals due to rounding
▲ Significantly higher than the average of countries
▼ Significantly lower than the average of countries

(24% of students) and least frequent in the Russian Federation (3%) and Slovenia (4%). An even higher percentage of grade four students (21% on average) indicated that a helium-filled balloon rising in the air is due to an upward push from gravity (item 14, not shown).[14] Both items 13 and 14 demonstrate that some grade four students have the misconception that gravity pushes upward on objects sitting on a solid surface or on objects that are moving upward.

The final two example items at grade four (items 15 and 16: Figs. 4.15 and 4.16) show that some grade four students attributed movement in a direction other than downward (e.g., horizontal movement) to the force of gravity. For item 15 (Fig. 4.15 and Table 4.15), about half of grade four students (51% on average

[14] Item 14 is a secure item from TIMSS 2015 and cannot be shown in the report.

across countries) correctly identified "a boy falling from a tree to the ground" (option D) as an example of an object moving because of the force of gravity. However, many students selected responses where an object moves in a direction other than downward (options A, B, and C). The percentage correct on this item ranged from 30–40% in Norway and Slovenia, to 53% in Italy, to >60% in the Russian Federation and the United States. The most common incorrect response in all countries was "a girl hitting a ball with a bat" (option A), which was selected by 21% of students on average. Similarly, for item 14,[15] 27% of students on average indicated that horizontal movement of objects was due to gravity.

In which example does an object move because of the force of gravity?

Ⓐ a girl hitting a ball with a bat
Ⓑ a boy pushing a box across the floor
Ⓒ a girl hammering a nail into a wall
Ⓓ a boy falling from a tree onto the ground

Item information
Item ID:
S041308

Year(s) administered:
2007

Performance objective:
Identify the direction of the force due to gravity

Correct answer:
D

Fig. 4.15 TIMSS grade four physics item 15, 2007. *Source* TIMSS 2007 Assessment. Copyright © 2009 International Association for the Evaluation of Educational Achievement (IEA). TIMSS & PIRLS International Study Center, Lynch School of Education, Boston College

Table 4.15 Student performance data for physics item 15 (S041308), 2007

Country	Percentage of students (%)						
	Correct (D)		A	B	C	D	Omitted
Italy	53		24	10	9	53	4
Norway	40	↓	29	17	9	40	5
Russian Federation	67	↑	11	11	7	67	4
Slovenia	33	↓	21	22	18	33	5
United States	61	↑	20	11	6	61	2
Average of countries	51		21	14	10	51	4

Notes Detail may not sum to totals due to rounding
↑ Significantly higher than the average of countries
↓ Significantly lower than the average of countries

[15] Item 14 is a secure item from TIMSS 2015 and cannot be shown in the report.

In item 16 (Fig. 4.16 and Table 4.16), many grade four students (43% on average) indicated that gravity can make objects "repel" (or move away from) each other (options A and C). The misconception was common in all five countries, ranging from 37% in the Russian Federation and 38% in Slovenia, to about 44% in Italy and the United States, to 54% of students in Norway. This demonstrates a lack of understanding at grade four that gravity is an attractive force that pulls objects toward Earth.

Which of the following can make objects repel each other?

Ⓐ gravity

Ⓑ magnetism

Ⓒ both gravity and magnetism

Ⓓ neither gravity nor magnetism

Item information
Item ID:
S031313

Year(s) administered:
2003

Performance objective:
Identify the direction of the force due to gravity

Correct answer:
B

Fig. 4.16 TIMSS grade four physics item 16, 2003. *Source* TIMSS 2003 Assessment. Copyright © 2005 International Association for the Evaluation of Educational Achievement (IEA). TIMSS & PIRLS International Study Center, Lynch School of Education, Boston College

Table 4.16 Student performance data for physics item 16 (S031313), 2003

Country	Percentage of students (%)						
	Correct (B)	A	B	C	D	Invalid	Omitted
Italy	39 ↑	21	39	23	11	1	5
Norway	27 ↓	31	27	23	10	1	8
Russian Federation	37	21	37	16	19	1	6
Slovenia	26 ↓	15	26	23	31	1	5
United States	45 ↑	13	45	31	8	0	3
Average of countries	35	20	35	23	16	1	5

Notes Detail may not sum to totals due to rounding
↑ Significantly higher than the average of countries
↓ Significantly lower than the average of countries

4.2.3 Patterns in Misconceptions and Misunderstandings Related to Gravity Across Grade Levels and Countries

Student performance data on the individual assessment items described in Sect. 4.2.2 were combined to explore patterns in the percentage of students demonstrating specific misconceptions, errors, and misunderstandings in each country based on the set of items that measure each type of misconception at each grade level (Figs. 4.17, 4.18, and 4.19).[16]

[16]The data shown in Figs. 4.17, 4.18, and 4.19 reflect the most recent assessment year, which differs across the set of items at each grade level (from 1995 to 2015). Table 4.2 shows the most recent assessment year for each item.

Misconception or misunderstanding	Percentage of students with misconception or misunderstanding (%)
Performance objective 1: Determine the acceleration of thrown objects (after they are released)	
P1A: Gravitational force (acceleration) acting on objects near Earth's surface is not constant but changes with the height of the object above the surface	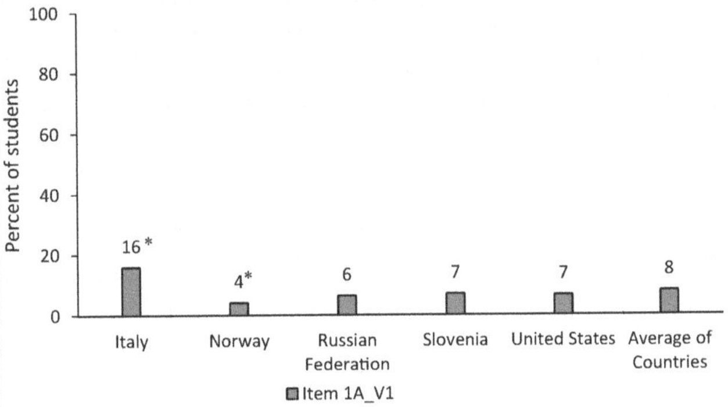
P1B: Objects thrown upward have no acceleration at their maximum height where the instantaneous velocity is zero (the instant it stops moving upward and reverses direction)	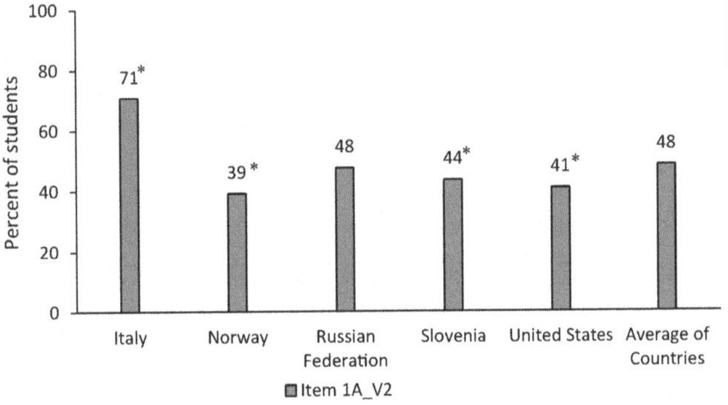
P1C: Gravitational acceleration is always in the direction of motion/velocity (rather than a constant acceleration directed toward the center of Earth)	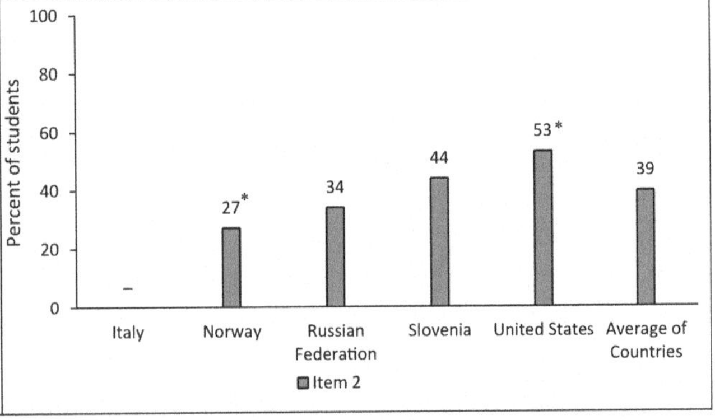

Fig. 4.17 Percentage of TIMSS Advanced students with misconceptions and misunderstandings about gravity, by country, 1995, 2008, and 2015. *Notes* Item 1A contributes to two misconceptions depending on the response options considered. For this item misconception P1A includes students who selected either option B or C, while misconception P1B includes students who selected option A. The percentages are for the most recent cycle each item was administered. Data for items 1A and 1B are from 2015; data for item 3 are from 2008, and data for item 2 are from 1995. *Significantly different from average of countries. – Data not available (see Appendix for country-specific notes)

<table>
<tr><th>Misconception or misunderstanding</th><th>Percentage of students with misconception or misunderstanding (%)</th></tr>
<tr><td colspan="2">Performance objective 2: Determine the time duration between different points on the path of a thrown object</td></tr>
<tr><td>P2: The time on the way up and the time on the way down are not equal (i.e., the downward acceleration due to gravity is not treated as constant)</td><td>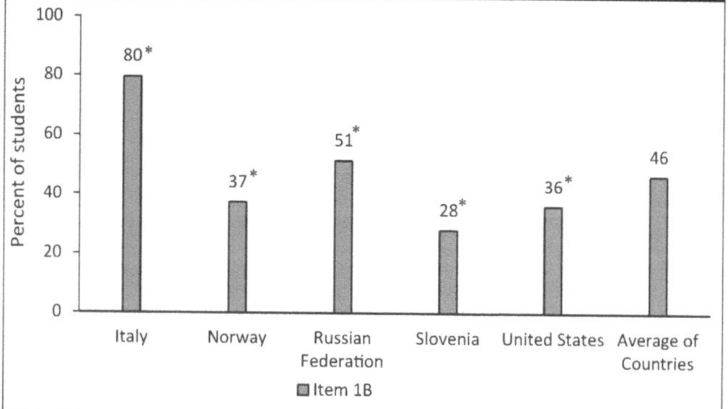
</td></tr>
<tr><td colspan="2">Performance objective 3: Determine the effect of gravitational force on moving objects or on objects at rest</td></tr>
<tr><td>P3A: Gravity acts only on falling objects, but not on objects at rest (on the ground or sitting on another surface) or on objects that are moving upward</td><td>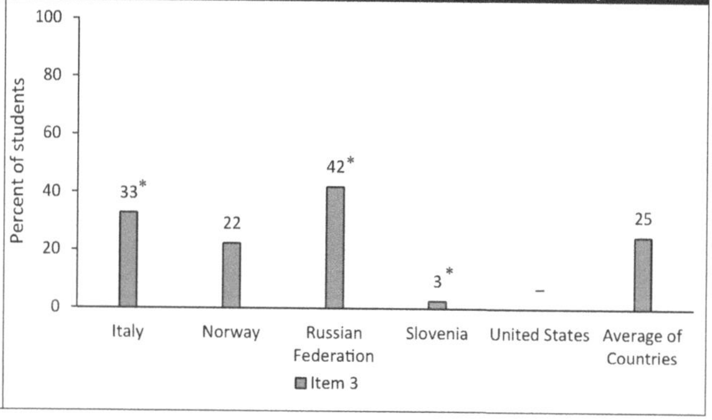
</td></tr>
</table>

Fig. 4.17 (continued)

All three of the TIMSS Advanced items were related to the motion of objects thrown upward or bouncing after being thrown. Both item 1A_V1 ("motion of a ball thrown upward—acceleration at highest point," Fig. 4.4) and item 2 ("acceleration of a bouncing ball," Fig. 4.5) measure performance objective 1: "determine the acceleration of thrown objects (after they are released)." Based on the frequency of specific response types (Fig. 4.17), many students in all five countries demonstrated the related misconceptions that "objects thrown upward have no acceleration at their maximum height where the instantaneous velocity is zero" (P1B) and that "gravitational acceleration is always in the direction of motion/velocity" (P1C). The other misconception measured by item 1A (V2, options B and C) was "gravitational force (acceleration) acting on objects near Earth's surface is not constant but changes with the height of the objects above the surface" (P1A). This was far less common, as many more students selected option A (acceleration is zero when the ball is at its maximum height). Across items, the percentage of students

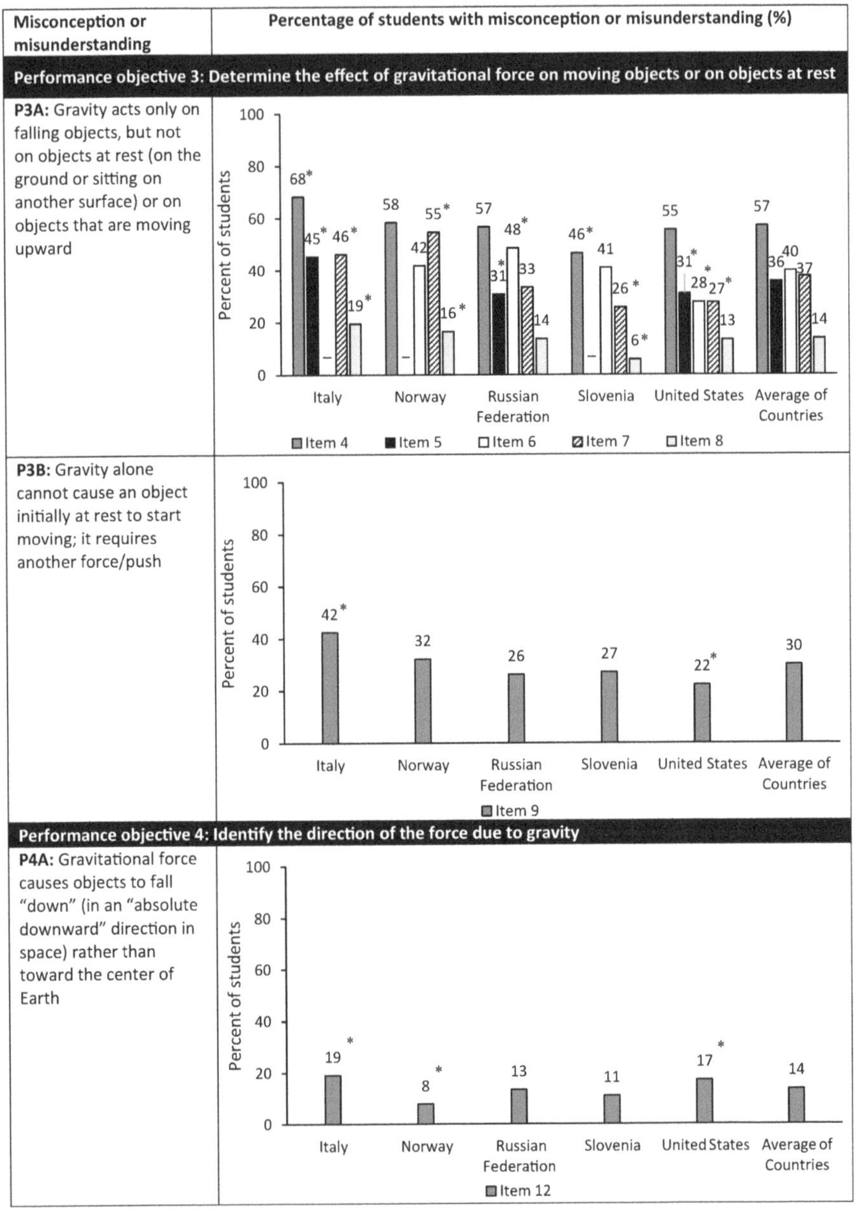

Fig. 4.18 Percentage of grade eight students with misconceptions and misunderstandings about gravity, by country, 1995, 1999, 2003, 2011, and 2015. *Notes* The percentages are for the most recent cycle each item was administered. Data for items 7 and 9 are from 2015; data for item 4 are from 2011; data for items 8 and 12 are from 2003; data for item 5 are from 1999; and data for item 6 are from 1995. *Significantly different from average of countries. – Data not available (see Appendix for country-specific notes)

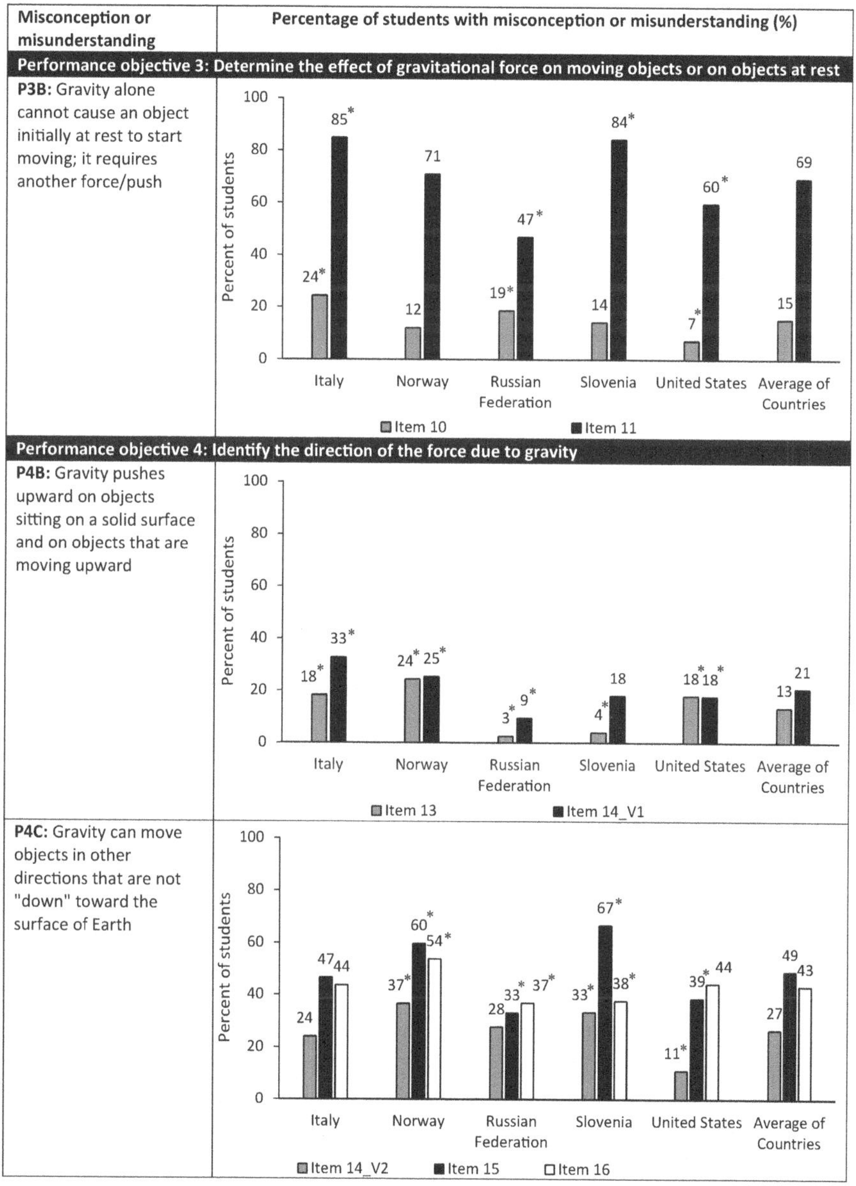

Fig. 4.19 Percentage of grade four students with misconceptions and misunderstandings about gravity, by country, 2003, 2007, 2011, and 2015. *Notes* Item 14 contributes to two misconceptions depending on the response options considered. For this item misconception P4B includes students who selected option C, while misconception P4C includes students who selected either option B or D. The percentages are for the most recent cycle each item was administered. Data for items 11, 13 and 14 are from 2015; data for item 10 are from 2011; data for item 15 are from 2007; and data for item 16 are from 2003. *Significantly different from the average of countries

demonstrating these types of misconceptions was higher in Italy and lower in Norway than the average across countries.

Performance objective 2 ("determine the time duration between different points on the path of a thrown object") was measured by item 1B ("motion of a ball thrown upward—time between two points"), and the misconception demonstrated (P2) was quite common in all countries. Item 3 ("forces acting on a stone thrown upward," Fig. 4.6) measures performance objective 3: "determine the effect of gravitational force on moving objects or on objects at rest." An incorrect response to this item illustrated a lack of understanding of how the force of gravity acts on objects when they are moving upward, which is related to misconception P3A. This misconception was more common than average in Italy and the Russian Federation and was very infrequent in Slovenia (3% of students). TIMSS Advanced students not identifying gravity in this item may have misconceptions commonly found at the lower grade levels that gravity does not act on objects while they are at rest or moving upward (see Figs. 4.18 and 4.19 for misconceptions about the effect of gravitational force in grade eight and grade four).

Considering the set of TIMSS Advanced items, we noted that the prevalence of specific types of misconceptions differed across countries. In Italy, Norway, and the Russian Federation, misconceptions P1B and P2 were more common, while in the United States, misconception P1C was most common; in Slovenia, misconceptions P1B and P1C were equally common.

The misconceptions held by TIMSS Advanced students that acceleration due to gravity is not constant can arise from related misconceptions about the force of gravity in earlier years. At grade eight (Fig. 4.18), six items measured performance objective 3: "determine the effect of gravitational force on moving objects or on objects at rest." Five of these items, namely item 4 (Fig. 4.7), item 5 (not shown), item 6 (not shown), item 7 (Fig. 4.8), and item 8 (Fig. 4.9), measured the misconception that "gravity acts only on falling objects, but not on objects at rest or on objects that are moving upward" (P3A). This misconception was very common across countries and, in particular, on item 4 ("gravity acting on a parachute jumper"), where >50% of all students demonstrated the misconception in every country except Slovenia (where this was 46%). The same misconception was measured in the similar item 5 ("gravity acting on a rocket being launched from Earth") and item 6 ("gravity acting on an apple falling from a tree"). Although the misconception did not appear to be as frequent on these two other items, it was still quite common in all countries, ranging from 31 to 45% on item 5 and from 28 to 48% on item 6. The misconception was also somewhat less common in item 7 ("forces acting on people sitting on a wall"), ranging from 26 to 55% of students. Only 6 to 19% of students demonstrated the misconception that "gravity does not act on objects that are moving upward" (P3A) in item 8 ("why helium balloon moves upward"). The different response patterns across these items in each country may be related to students' familiarity with the specific contexts. In general, the frequency of misconception P3A at grade eight was less in Slovenia than on average across countries, which was also the case for the TIMSS Advanced item 3, which measured the same type of misconception (Fig. 4.17).

Grade eight item 9 ("force causing a ball thrown upward to fall," Fig. 4.10), measured the misconception that "gravity alone cannot cause an object initially at rest to start moving; it requires another force/push" (P3B). This misconception, which was common at grade four (Fig. 4.19), was also demonstrated by many grade eight students for this item (Fig. 4.18), with the frequency ranging from 22% of students in the United States to 42% of students in Italy. On average, 30% of grade eight students demonstrated this misconception. This is similar to the average percentage of TIMSS Advanced students on item 3 (25%) (Fig. 4.17) who demonstrated the related misconception P3A by not identifying gravity as a force that acts on a stone after it is thrown straight up in the air. However, there was less variation across countries in the percentage of grade eight students demonstrating the misconception than in the percentage of TIMSS Advanced students.

The last grade eight item, item 12 ("direction gravity makes a ball fall at different places on Earth," Fig. 4.11), measured performance objective 4: "identify the direction of the force due to gravity." The misconception demonstrated on this item that "gravitational force causes objects to fall down (in an absolute downward direction in space) rather than toward the center of Earth" (P4A) was less common than the other two types of misconceptions demonstrated by students at grade eight. Misconceptions or misunderstandings related to the direction of the force of gravity, however, were quite common at grade four (Fig. 4.19).

At grade four, two items (10 and 11) measure performance objective 3: "determine the effect of gravitational force on moving objects or on objects at rest." In both items, incorrect responses demonstrated the misconception that "gravity alone cannot cause an object initially at rest to start moving; it requires another force/push" (P3B), but the frequency of the misconception is quite different on the two items. The misconception was demonstrated by 15% of students on average on MC item 10 ("force causing an object to fall to the ground," Fig. 4.12), compared to 69% on average on CR item 11 ("force causing a marble to roll down a sloping track," Fig. 4.13). This illustrates that while most students at grade four demonstrated basic knowledge about the role of gravity in falling objects (item 10), many could not apply this in a less familiar context by connecting gravity to the motion of an object rolling down a sloped surface (item 11).

The remaining grade four items measure performance objective 4: "identify the direction of the force due to gravity." Item 13 ("direction of the force of Earth's gravity," Fig. 4.14) and item 14_V1 ("direction of movement due to gravity," not shown) measure the misconception that "gravity pushes upward on objects sitting on a solid surface and on objects that are moving upward" (P4B). This misconception was most common in Norway and Italy, and least common in the Russian Federation. Other grade four items measure the related misconception that "gravity can move objects in other directions that are not 'down' toward the surface of Earth" (P4C). Item 14_V2 ("direction of movement due to gravity"), item 15 ("example of an object moving due to gravity," Fig. 4.15), and item 16 ("force that makes objects repel each other," Fig. 4.16) all measured the misconception that gravity can make objects move in a horizontal direction, and demonstrated a lack of understanding among grade four students that gravity is an attractive force that pulls objects toward the surface of the Earth. The misconceptions on these items were generally common, ranging from 27 to 49% on average. Like misconception P4B, misconception P4C was most common in Norway. This lack of basic understanding at grade four can lead to misconceptions and misunderstandings at higher grade levels, such as those illustrated by items 7 and 12 at grade eight (Fig. 4.18) and by item 2 in TIMSS Advanced (Fig. 4.17).

Across the set of items at grade four, gravity misconceptions were frequently less common than average in the Russian Federation and the United States, and more common than average in Italy, Norway, and Slovenia. This pattern, however, did not persist across the misconceptions at higher grade levels. In particular, the percentages of TIMSS Advanced students demonstrating the gravity misconceptions were generally lower in Norway and higher in the Russian Federation (or not measurably different from the average across countries). The patterns at grade eight were more mixed, with some countries having a higher frequency of some misconceptions and a lower frequency of others. The percentage of students demonstrating gravity misconceptions in Italy was higher than the average for all countries at all three grade levels that were assessed by TIMSS.

4.2.4 Gender Differences in Misconceptions and Misunderstandings Related to Gravity

On average across the five countries, male students outperformed female students on nearly all of the gravity items at all three grade levels (Table 4.17).[17] The only item where there were no significant gender differences in the percentage of students who were correct in any country was grade eight item 12 ("direction gravity makes a ball fall at different places on Earth"). Patterns in performance by gender differed across countries and grades. Gender differences in the percent correct were greatest on the TIMSS Advanced items, with an average male–female difference of at least 10%. In comparison, the average male–female difference in the percent correct ranged from 6 to 12% at grade eight and from 2 to 13% at grade four. In TIMSS Advanced, gender differences were pervasive in Norway, with a significantly higher percentage of males than females responding correctly to all items. In contrast, in Norway, none of the grade eight items, and three of seven items at grade four showed significant gender differences. By comparison, in the United States, there were significant gender differences on all of the grade four items, three of seven items at grade eight, and two of three items in TIMSS Advanced. In Italy, the Russian Federation, and Slovenia, significant gender differences were found for one to three items at each grade level. The specific set of items with significant gender differences varied across countries. None of the items demonstrated

[17]Table 4.17 displays the percent correct for female and male students on each physics item. The corresponding percent correct of students overall are shown in Figs. 4.1, 4.2, and 4.3.

Table 4.17 Performance of female and male students on TIMSS and TIMSS Advanced physics items, by country and grade level, 1995–2015

Item	Year	Percentage of students correct (%)											
		Italy		Norway		Russian Federation		Slovenia		United States		Average of countries	
		Female	Male	Female	Male	Female	Male	Female	Male	Female	Male	Female	Male
TIMSS Advanced													
Item 1A	2015	7	11	40	63	41	49	37	53	35	61	32	47
Item 1B	2015	14	26	52	67	39	55	67	74	55	69	45	58
Item 2	1995	–	–	26	52	20	25	8	17	9	18	16	28
Item 3	2008	27	35	38	59	37	47	58	64	–	–	40	51
TIMSS grade 8													
Item 4	2011	20	32	28	36	33	44	48	45	34	40	33	40
Item 5	1999	19	30	–	–	39	54	–	–	39	53	32	46
Item 6	1995	–	–	45	54	38	51	51	55	65	64	50	56
Item 7	2015	43	64	42	50	63	71	75	74	72	73	59	66
Item 8	2003	58	63	67	71	72	78	75	83	67	78	68	74
Item 9	2015	54	61	65	70	74	73	69	77	77	79	68	72
Item 12	2003	72	70	84	85	82	84	87	80	73	77	79	79
TIMSS grade 4													
Item 10	2011	54	67	63	70	59	65	55	58	76	87	61	69
Item 11	2015	12	18	23	35	52	55	16	16	35	46	27	34
Item 13	2015	69	74	62	63	94	96	87	89	67	77	76	80
Item 14	2015	33	42	25	39	55	66	29	50	61	74	41	54
Item 15	2007	53	54	41	40	64	69	30	37	58	65	49	53
Item 16	2003	40	39	26	29	35	40	27	24	41	49	34	36

Key

- (dark shading) Higher percentage (%) of females with item correct
- (light shading) Higher percentage (%) of males with item correct
- (no shading) No significant difference between females and males

Notes Percent correct is the percentage of students receiving credit on each item. For MC and short CR items (each worth one score point), this reflects the percentage of students who provided a correct answer. For extended CR items, this reflects the weighted percentage of students receiving full credit (two points) or partial credit (one point)

– Data not available (see Appendix for country-specific notes)

significant gender differences in all five countries, but there were two items at each grade level with significant gender differences in three or four countries.

To further understand these gender differences in item performance within and across countries, we compared the percentage of male and female students demonstrating specific types of misconceptions and misunderstandings at each grade level (Tables 4.18, 4.19, and 4.20, and Figs. 4.20, 4.21, and 4.22).[18]

In TIMSS Advanced (Table 4.18 and Fig. 4.20), a higher percentage of female students on average demonstrated misconceptions P1A and P1B on item 1A ("motion of a ball thrown upward—acceleration at highest point") and misconception P2 on item 1B ("motion of a ball thrown upward—time between two points"). The percentage differences were greatest for misconception P1B, where between 16 and 20% more females than males demonstrated the misconception in Norway, Slovenia, and the United States. In contrast, in Italy 9% more males demonstrated misconception P1B, while 9% more females demonstrated misconception P1A. In the Russian Federation, there was no significant difference in the percentage of males and females demonstrating misconception P1B, but there were 6% more females demonstrating misconception P1A. For misconception P2, between 13 and 16% more females demonstrated the misconception in Italy, Norway, the Russian Federation, and the United States. There were no significant differences found in the percentages of male and female students demonstrating misconception P1C on item 2 ("acceleration of a bouncing ball") or misconception P3A on item 3 ("forces acting on a stone thrown upward").

Compared to TIMSS Advanced, there were fewer significant differences in the percentages of male and female students demonstrating the misconceptions at grade eight (Table 4.19 and Fig. 4.21). Across the five items measuring misconception P3A ("gravity acts only on falling objects, but not on objects at rest or on objects that are moving upward"), there were two items in Italy and the Russian Federation, and only one item in the United States, where a significantly higher percentage of females demonstrated the misconception. In Norway and Slovenia, none of these items showed significant gender differences. The specific set of items with a higher percentage of female students having the misconception varied across countries. On item 4 ("gravity acting on a parachute jumper"), 10% more females than males demonstrated the misconception in both Italy and the Russian Federation, while on item 5 ("gravity acting on a rocket being launched from Earth"), only the Russian Federation showed a gender difference (13% more female than male students demonstrating the misconception). There were no significant gender differences on item 6 ("gravity acting on an apple falling from a tree"). The largest gender difference was seen in Italy for item 7 ("forces acting on people sitting on a wall"), where 21% more female than male students demonstrated the misconception. In the

[18]Tables 4.18, 4.19 and 4.20 display the percentage of female and male students with each misconception in TIMSS Advanced, grade eight and grade four, respectively. The accompanying figures (Figs. 4.20, 4.21, and 4.22) illustrate the differences in the percentages of female and male students at the corresponding grade levels. The corresponding overall percentages of students with the misconceptions are shown in Figs. 4.17, 4.18, and 4.19.

Table 4.18 Percentage of female and male TIMSS Advanced students with misconceptions and misunderstandings about gravity, by country, 1995, 2008, and 2015

Misconceptions and misunderstandings	Item	Year	Percentage of students with misconception or misunderstanding (%)											
			Italy		Norway		Russian Federation		Slovenia		United States		Average of countries	
			Female	Male	Female	Male	Female	Male	Female	Male	Female	Male	Female	Male
Performance objective 1: Determine the acceleration of thrown objects (after they are released)														
P1A: Gravitational force (acceleration) acting on objects near Earth's surface is not constant but changes with the height of the object above the surface	Item 1A _V1	2015	20	12	6	3	10	4	5	8	13	3	11	6
P1B: Objects thrown upward have no acceleration at their maximum height where the instantaneous velocity is zero (the instant it stops moving upward and reverses direction)	Item 1A _V2	2015	66	75	53	33	48	47	56	38	51	35	55	46
P1C: Gravitational acceleration is always in the direction of motion/velocity (rather than a constant acceleration directed toward the center of Earth)	Item 2	1995	–	–	34	25	35	33	48	43	49	55	42	39
Performance objective 2: Determine the time duration between different points on the path of a thrown object														
P2: The time on the way up and the time on the way down are not equal (i.e., the downward acceleration due to gravity is not treated as constant)	Item 1B	2015	86	74	48	33	61	45	33	26	45	31	55	42
Performance objective 3: Determine the effect of gravitational force on moving objects or on objects at rest														
P3A: Gravity acts only on falling objects, but not on objects at rest (on the ground or sitting on another surface) or on objects that are moving upward	Item 3	2008	29	36	28	20	45	40	3	2	–	–	26	24

Key

- Higher percentage (%) of females with error or misunderstanding
- Higher percentage (%) of males with error or misunderstanding
- No significant difference between females and males

Notes Item 1A contributes to two misconceptions depending on the response options considered. For this item misconception P1A includes students who selected either option B or C, while misconception P1B includes students who selected option A

– Data not available (see Appendix for country-specific notes)

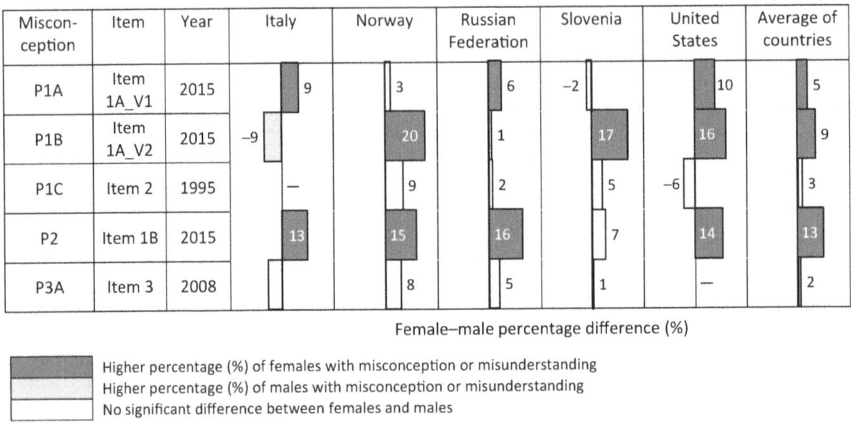

Fig. 4.20 Gender differences in misconceptions and misunderstandings about gravity among TIMSS Advanced students, 1995, 2008, and 2015. *Notes* Physics misconceptions and misunderstandings: P1A = gravitational force (acceleration) acting on objects near Earth's surface is not constant but changes with the height of the object above the surface, P1B = objects thrown upward have no acceleration at their maximum height where the instantaneous velocity is zero (the instant it stops moving upward and reverses direction), P1C = gravitational acceleration is always in the direction of motion/velocity (rather than a constant acceleration directed toward the center of Earth), P2 = the time on the way up and the time on the way down are not equal (the downward acceleration due to gravity is not treated as constant), P3A = gravity acts only on falling objects, but not on objects at rest (on the ground or sitting on another surface) or on objects that are moving upward. Item 1A contributes to two misconceptions depending on the response options considered. For this item, misconception P1A includes students who selected either option B or C, while misconception P1B includes students who selected option A. – Data not available (see Appendix for country-specific notes)

United States, the only significant gender difference was on item 8 ("why helium balloon moves upward"), where 8% more female than male students demonstrated the misconception. The only significant gender difference on item 9 ("force causing a ball thrown upward to fall") was in Slovenia, where 9% more females than males demonstrated misconception P3B ("gravity alone cannot cause an object initially at rest to start moving; it requires another force/push"). There were no significant differences in the percentages of male and female students demonstrating misconception P4A on item 12 ("direction gravity makes a ball fall at different places on Earth").

As in grade eight and TIMSS Advanced, there were different patterns of gender differences in misconceptions and misunderstandings across countries on the grade four items (Table 4.20 and Fig. 4.22). Most notably, only in the United States was the frequency of misconceptions and misunderstandings significantly higher for females than males on all items, with the differences ranging from 5 to 11%. Across the two items measuring misconception P3B, significantly more females than males in Italy (11% on item 10), Norway (11% on item 11), and the United States (5% on item 10 and 11% on item 11) demonstrated the misconception that "gravity alone

Table 4.19 Percentage of female and male grade eight students with misconceptions and misunderstandings about gravity, by country, 1995, 1999, 2003, 2011, and 2015

Misconceptions and misunderstandings	Item	Year	Percentage of students with misconception or misunderstanding (%)											
			Italy		Norway		Russian Federation		Slovenia		United States		Average of countries	
			Female	Male	Female	Male	Female	Male	Female	Male	Female	Male	Female	Male
Performance objective 3: Determine the effect of gravitational force on moving objects or on objects at rest														
P3A: Gravity acts only on falling objects, but not on objects at rest (on the ground or sitting on another surface) or on objects that are moving upward	Item 4	2011	73	63	61	55	61	51	47	46	57	54	60	54
	Item 5	1999	43	47	–	–	37	24	–	–	34	28	38	33
	Item 6	1995	–	–	42	41	53	42	43	39	28	27	42	37
	Item 7	2015	57	36	58	50	37	29	25	26	28	27	41	34
	Item 8	2003	18	21	16	17	15	12	6	5	17	9	14	13
P3B: Gravity alone cannot cause an object initially at rest to start moving; it requires another force/push	Item 9	2015	46	39	35	30	26	27	31	23	23	21	32	28
Performance objective 4: Identify the direction of the force due to gravity														
P4A: Gravitational force causes objects to fall "down" (in an "absolute downward" direction in space) rather than toward the center of Earth	Item 12	2003	17	21	9	7	17	10	8	14	19	16	14	14

Key

- (dark shading) Higher percentage (%) of females with error or misunderstanding
- (light shading) Higher percentage (%) of males with error or misunderstanding
- (no shading) No significant difference between females and males

Notes

– Data not available (see Appendix for country-specific notes)

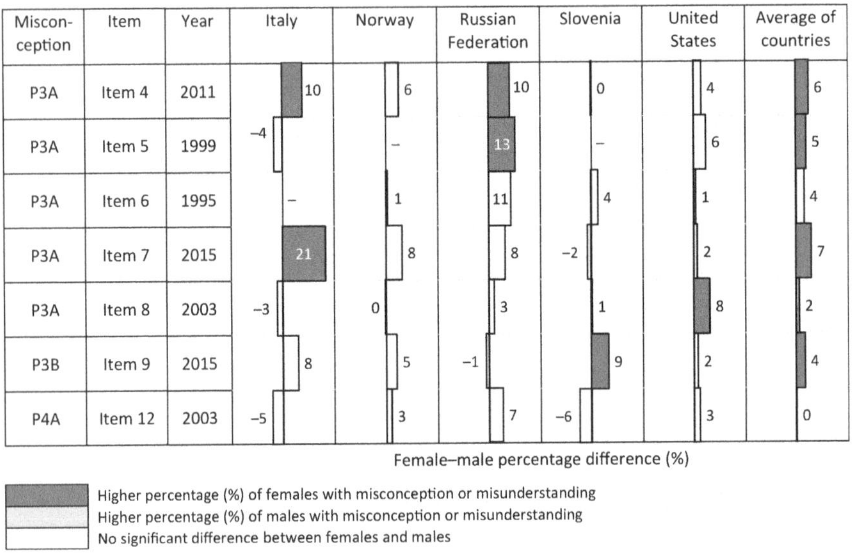

Fig. 4.21 Gender differences in misconceptions and misunderstandings about gravity among grade eight students, 1995, 1999, 2003, 2011, and 2015. *Notes* Physics misconceptions and misunderstandings: P3A = gravity acts only on falling objects, but not on objects at rest (on the ground or sitting on another surface) or on objects that are moving upward, P3B = gravity alone cannot cause an object initially at rest to start moving; it requires another force/push, P4A = gravitational force causes objects to fall "down" (in an "absolute downward" direction in space) rather than toward the center of Earth. – Data not available (see Appendix for country-specific notes)

cannot cause an object initially at rest to start moving; it requires another force/push." On both items measuring misconception P4B ("gravity pushes upward on objects sitting on a solid surface and objects that are moving upward"), only the United States had significantly higher percentages of females demonstrating the misconception (7–8%). The largest gender differences were found on item 14_V2 ("direction of movement due to gravity"), where the percentage of females was significantly higher than the percentage of males demonstrating misconception P4C ("gravity can make objects move in other directions that are not down toward the surface of the Earth") in Norway (18%), the Russian Federation (9%), Slovenia (17%), and the United States (7%).

Table 4.20 Percentage of female and male grade four students with misconceptions and misunderstandings about gravity, by country: 2003, 2007, 2011, and 2015

Misconceptions and misunderstandings	Item	Year	Percentage of students with misconception or misunderstanding (%)											
			Italy		Norway		Russian Federation		Slovenia		United States		Average of countries	
			Female	Male	Female	Male	Female	Male	Female	Male	Female	Male	Female	Male
Performance objective 3: Determine the effect of gravitational force on moving objects or on objects at rest														
P3B: Gravity alone cannot cause an object initially at rest to start moving; it requires another force/push	Item 10	2011	30	19	15	9	21	16	15	13	10	5	18	13
	Item 11	2015	88	82	77	65	48	45	84	84	65	54	73	66
Performance objective 4: Identify the direction of the force due to gravity														
P4B: Gravity pushes upward on objects sitting on a solid surface and on objects that are moving upward	Item 13	2015	20	17	24	25	3	2	4	4	21	14	14	12
	Item 14_V1	2015	36	30	24	27	11	8	19	17	22	14	22	19
P4C. Gravity can make objects move in other directions that are not "down" toward the surface of Earth	Item 14_V2	2015	25	23	46	28	33	23	42	25	14	8	32	21
	Item 15	2007	47	46	59	60	36	31	70	63	42	35	51	47
	Item 16	2003	44	44	54	53	37	37	40	36	47	41	44	42

Key

- Higher percentage (%) of females with error or misunderstanding (dark shading)
- Higher percentage (%) of males with error or misunderstanding (light shading)
- No significant difference between females and males (no shading)

Notes Item 14 contributes to two misconceptions depending on the response options considered. For this item, misconception P4B includes students who selected option C, while misconception P4C includes students who selected either option B or D

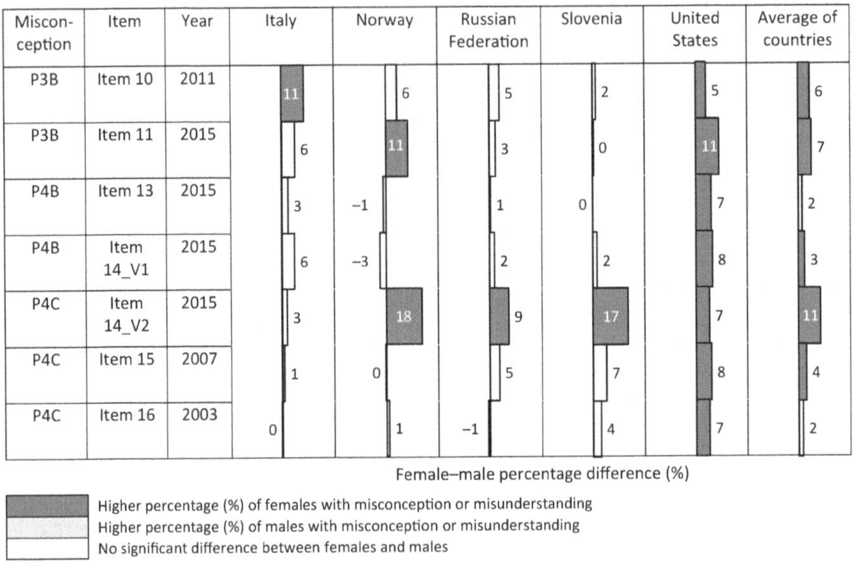

Fig. 4.22 Gender differences in misconceptions and misunderstandings about gravity among grade four students, 2003, 2007, 2011, and 2015. *Notes* Physics misconceptions and misunderstandings: P3B = gravity alone cannot cause an object initially at rest to start moving; it requires another force/push, P4B = gravity pushes upward on objects sitting on a solid surface and on objects that are moving upward, P4C = gravity can move objects in other directions that are not "down" toward the surface of Earth. Item 14 contributes to two misconceptions depending on the response options considered. For this item, misconception P4B includes students who selected option C, while misconception P4C includes students who selected either option B or D

4.2.5 *Patterns in Misconceptions and Misunderstandings Related to Gravity Over Time*

In this section, we present the percentage of students in each country demonstrating a specific type of misconception or misunderstanding over multiple assessment years for the set of trend items from each grade level (Fig. 4.23 and 4.24). For the gravity topic, there were three trend items each in grades four and eight, but no trend items available for TIMSS Advanced. All items except one at grade four (item 11) were administered in three assessment years before they were released.

At grade eight (Fig. 4.23), the trend item data covered assessment years 2003, 2007, 2011, and 2015. There were some significant differences across assessment years in the percentage of students in each country demonstrating the specific types of misconceptions. Item 4 (administered in 2003, 2007, and 2011) and item 7 (administered in 2007, 2011, and 2015) both measure misconception P3A ("gravity acts only on falling objects, but not on objects at rest or on objects that are moving upward"). For MC item 4 ("gravity acting on a parachute jumper"), the percentage of students demonstrating the misconception was not measurably different in 2011 than in the previous two assessments in Italy, Norway, the Russian Federation, and the United States. In contrast, in Slovenia, the percentage of students decreased between 2003 (55%) and 2011 (46%). For CR item 7 ("forces acting on people

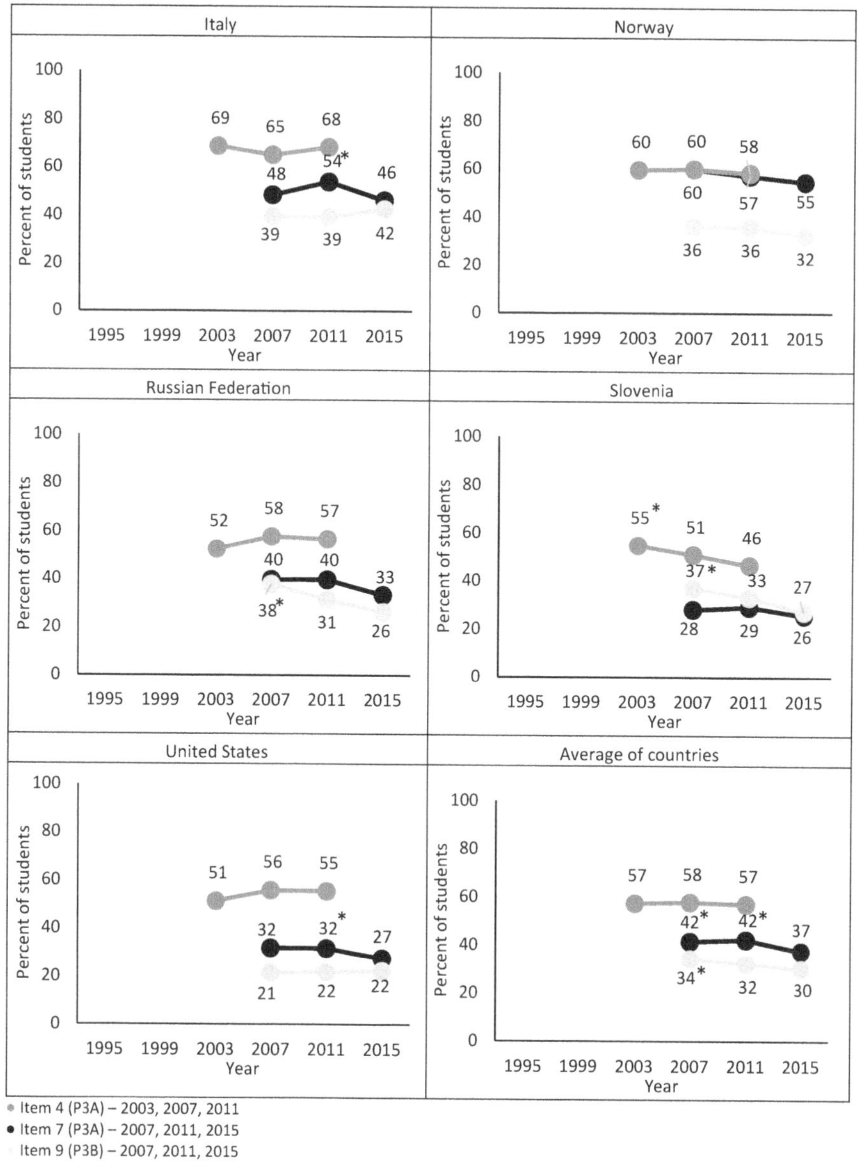

Fig. 4.23 Trends in the percentage of grade eight students with misconceptions and misunderstandings about gravity, 2003–2015. *Notes* Physics misconceptions and misunderstandings: P3A = gravity acts only on falling objects, but not on objects at rest (on the ground or sitting on another surface) or on objects that are moving upward, P3B = gravity alone cannot cause an object initially at rest to start moving; it requires another force/push. *Significantly different from most recent assessment cycle

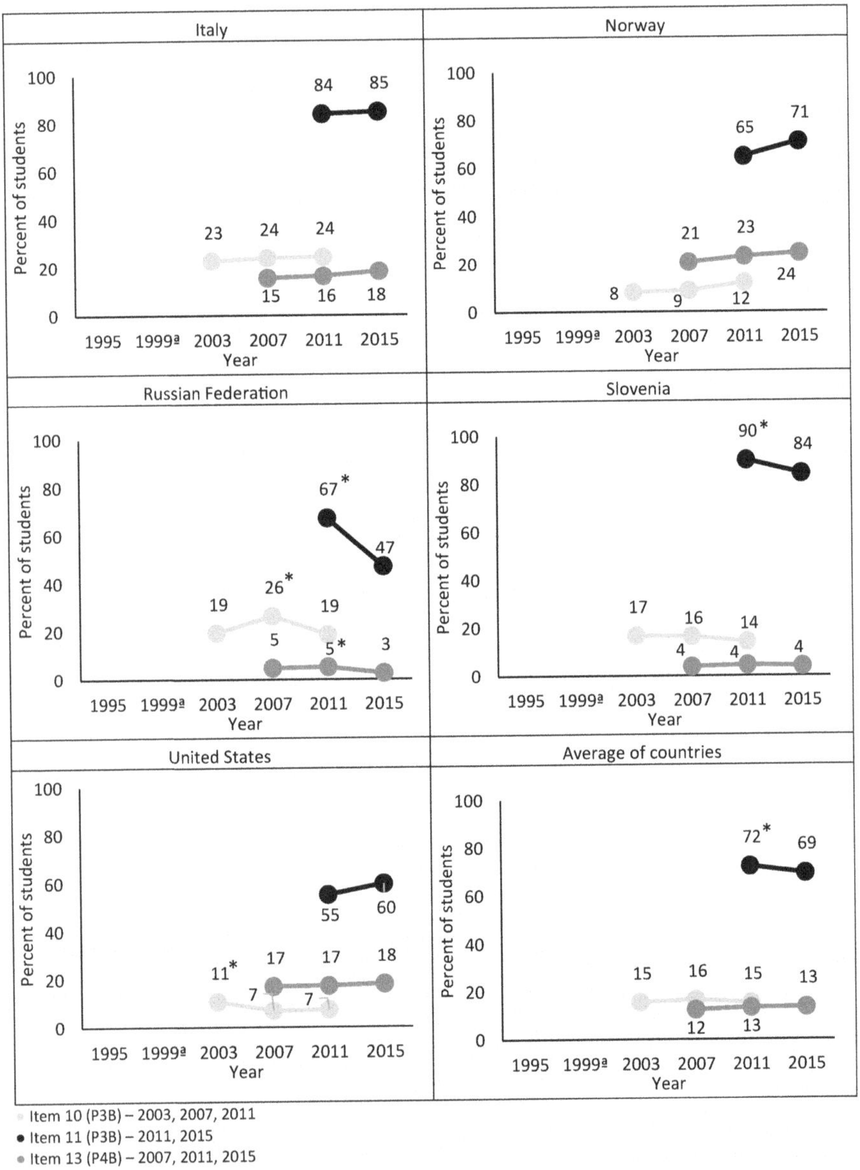

Fig. 4.24 Trends in the percentage of grade four students with misconceptions and misunderstandings about gravity, 2003–2015. *Notes* Physics misconceptions and misunderstandings: P3B = gravity alone cannot cause an object initially at rest to start moving; it requires another force/push, P4B = gravity pushes upward on objects sitting on a solid surface and on objects that are moving upward. *Significantly different from most recent assessment cycle. [a]TIMSS was not administered in 1999 at grade four

sitting on a wall"), there were significant differences in Italy and the United States but not in the other countries. In both Italy and the United States, the percentage of students with the misconception in 2011, the second assessment cycle, was higher than in 2015, while the percentage in 2007 was not measurably different. This reflected an 8% decrease in the percentage of students in Italy and a 5% decrease in the United States between 2011 and 2015. The largest differences were found for item 9 ("force causing a ball thrown upward to fall"), which measured misconception P3B ("gravity alone cannot cause an object initially at rest to start moving; it requires another force/push"). In both the Russian Federation and Slovenia, the percentage of students demonstrating the misconception decreased over time, and the difference between the first assessment cycle (2007) and the third (2015) was statistically significant. In Slovenia, this reflected a 10% decrease (from 37 to 27%) and in the Russian Federation, a 12% decrease (from 38 to 26%), with no measurable differences in Italy, Norway, and the United States.

At grade four (Fig. 4.24), the trend item data also covered assessment years 2003, 2007, 2011, and 2015. There were fewer significant differences over time in the percentage of students with misconceptions on the grade four items than at grade eight. As at grade eight, however, the most substantial differences were found in the Russian Federation and Slovenia. Items 10 and 11 both measure misconception P3B ("gravity alone cannot cause an object initially at rest to start moving; it requires another force/push"). On MC item 10 ("force causing object to fall to the ground"), the only countries with a significant difference were the Russian Federation and the United States. In the Russian Federation, the percentage of students demonstrating the misconception was 7% lower in 2011 (19%) than in 2007 (26%). In the United States, the percentage was lower in 2011 (7%) than in 2003 (11%), although the decrease appears to have occurred between 2003 and 2007 (also 7%). On CR item 11 ("force causing a marble to roll down a sloping track"), the percentage of students demonstrating the misconception decreased significantly in the Russian Federation (from 67 to 47%) and Slovenia (from 90 to 84%) between the 2011 and 2015 assessments. Item 13 ("direction of the force of Earth's gravity") measures misconception P4B ("gravity pushes upward on objects sitting on a solid surface and on objects that are moving upward"). There was only one statistically significant difference on this item, and that was a slight decrease of 2% of students in the Russian Federation between the 2011 and 2015 assessments (from 5% to 3%).

4.2.6 *Summary of Physics Results*

In the physics results sections, we have reported on students' performance on the set of items related to gravity across countries at each grade level (TIMSS Advanced, grade eight, and grade four; see Sect. 4.2.1), patterns in student misconceptions and misunderstandings across countries and grade levels (Sects. 4.2.2 and 4.2.3), gender differences in these misconceptions and misunderstandings (Sect. 4.2.4), and trends over multiple assessment years (Sect. 4.2.5). The frequency of specific types of student misconceptions and misunderstandings at each grade level varied across the five countries included in the study. In each country, and at each grade level, there were some misconceptions and misunderstandings that were demonstrated by at least one-third of students. Gender differences were found at all three grade levels, but were most prevalent on the TIMSS Advanced items. For all trend items (except one in TIMSS Advanced), gender differences favored males, with higher percentages of female students than male students demonstrating the misconception or misunderstanding. Performance on grade eight and grade four trend items administered in multiple assessment years showed that the frequency of certain student misconceptions and misunderstandings decreased over time in some countries but not others.

4.3 Mathematics Results

We selected a set of 29 mathematics items from the TIMSS and TIMSS Advanced assessments from 1995 to 2015 that measured student understandings and errors related to linear equations. This item set included two items from TIMSS Advanced,[19] 19 TIMSS items at grade eight, and eight TIMSS items at grade four. We identified nine performance objectives (POs) related to linear equations that were measured by these items, each with a specific set of related errors and misunderstandings (Table 4.21).

We provide here a list of the full set of TIMSS and TIMSS Advanced items related to linear equations (Table 4.22) organized by performance objective and grade level. This list shows the assessment year(s) when each item was administered, the item format (MC or CR), a brief item description, the figures where the items are shown in the report (released items only), and the specific type(s) of student errors and misunderstandings measured by each item. All mathematics results reported in this section are based on student performance on this set of items. (See Appendix Table A.2 for additional information on the mathematics items used in this study, including the specific response options or score categories used to determine the percentage of students demonstrating each type of error or misunderstanding.)

[19]Parts A and B of TIMSS Advanced item 1 are treated as separate items in this report (items 1A and 1B).

Table 4.21 Mathematics performance objectives related to linear equations with related errors and misunderstandings, by grade level

Performance objective	Related errors and/or misunderstandings	Grade level		
		TA	Gr8	Gr4
MO1: Interpret the solution to a system of linear equations to answer a question or solve a problem in real life	Not able to use slope and intercept to provide an argument in support of the solution to a real-life problem situation (M1)	✓		
MO2: Solve systems of linear equations in two variables	Not able to apply the procedure correctly to solve a real-life problem situation (M2A)	✓	✓	
	Not able to apply the procedure correctly to solve non-contextualized problems (M2B)		✓	
MO3: Interpret the meanings of slope and y-intercept in linear equations or graphs	Not able to relate slope with steepness of a line. (M3A)		✓	
	Demonstrates confusion between slope and intercept of an equation. (M3B)		✓	
MO4: Relate algebraic equations to their graphical representations (and vice versa)	Not able to correctly identify the graph of an equation (M4A)		✓	
	Not able to translate graphical representations into a mathematical equation or verbal description of a linear relationship (M4B)		✓	
MO5: Write equations to represent situations	Not able to translate verbal descriptions into a correct mathematical equation (M5)		✓	
MO6. Given pairs of numbers in tables or ordered pairs, generate an algebraic equation of the relationship between two variables	Not able to translate relationship shown in table form into a mathematical equation (M6)		✓	
MO7 Given pairs of numbers in tables or ordered pairs, generate a verbal description of the relationship	Not able to generate a correct verbal description given a specific relationship in the form of ordered pairs (M7A)		✓	
	Not able to generate a correct verbal description given a specific relationship shown in table form (M7B)			✓
MO8: Give a verbal description of a relationship between a set of numbers, generate pairs of whole numbers that follow that relationship (rule)	Not able to identify a correct set of numbers that follow a given relationship/rule (M8)			✓
MO9: Apply algebraic thinking to solve simple real-life problems involving unknowns	Not able to apply algebraic thinking to solve simple real-life problems involving unknowns (M9)			✓

Notes The nine mathematics performance objectives (MO1 through MO9). The related errors and misunderstandings are coded (e.g., M1, M2A, M2B, etc.). The first two identifiers refer to the corresponding mathematics objective number (e.g., M1, M2, etc.). When there is more than one error or misunderstanding under a performance objective, a third identifier was added (i.e., A, B, C). Grade levels: TA = TIMSS Advanced, G8 = grade 8, G4 = grade 4

✓ Indicates that the error or misunderstanding was measured by one or more items at that grade level

Table 4.22 List of TIMSS and TIMSS Advanced mathematics items related to linear equations, organized by performance objective and grade level

Item	Grade level	Assessment year(s)	Item format	Item description	Figure	Mathematics errors and misunderstandings												
						M1	M2A	M2B	M3A	M3B	M4A	M4B	M5	M6	M7A	M7B	M8	M9
Performance objective 1: Interpret the solution to a system of linear equations to answer a question or solve a problem in real life contexts																		
Item 1B (MA33240B)	TIMSS Advanced	2015	CR	Compare car rental plans X and Y—interpret solution	4.28	✓												
Performance objective 2: Solve systems of linear equations in two variables																		
Item 1A (MA33240A)	TIMSS Advanced	2015	CR	Compare car rental plans X and Y—solve system of equations	4.28		✓											
Item 2 (M042263)	Grade 8	2007	CR	Cost of 1 pen and 2 pencils	4.29		✓											
Item 3 (M062237)		2015	CR	Solve system of equations for x and y	4.30			✓										
Item 4 (M052087)[a]		2011, 2015	CR	Solve for simultaneous equations	–			✓										
Item 5 (M032728)		2003	MC	If x plus 3y equals 11 and 2× plus 3y equals 13	–			✓										
Performance objective 3: Interpret the meanings of slope and y-intercept in linear equations or graphs																		
Item 6 (M052105)[a]	Grade 8	2011, 2015	CR	Which slope has a greater value	–				✓									
Item 7 (M012022)		1995, 1999	MC	Equation to determine cost of cards	–					✓								

(continued)

Table 4.22 (continued)

Item	Grade level	Assessment year(s)	Item format	Item description	Figure	Mathematics errors and misunderstandings												
						M1	M2A	M2B	M3A	M3B	M4A	M4B	M5	M6	M7A	M7B	M8	M9
Performance objective 3: Interpret the meanings of slope and y-intercept in linear equations or graphs																		
Item 8 (M062095)[a]	Grade 8	2015	MC	Equation for Bruno's monthly cell phone cost	–					✓								
Item 9 (M062074)		2015	MC	Function for Janet's graph given slope/intercept	4.31					✓								
Performance objective 4: Relate algebraic equations to their graphical representations (and vice versa)																		
Item 10 (M052092)[a]	Grade 8	2011, 2015	MC	Graph of y = 2x	–						✓							
Item 11 (M042112)		2007, 2011, 2015	MC	Relationship between x and y	4.32							✓						
Item 12 (M062242)[a]		2015	MC	Relationship a graph of a line in words	–							✓						
Performance objective 5: Write equations to represent situations																		
Item 13 (M042202)	Grade 8	2007, 2011, 2015	MC	Formula for K the cost of trip	4.33								✓					
Item 14 (M062241)[a]		2015	CR	Roy's phone business—equation for y	–								✓					
Item 15 (M032545)		2003	CR	7 oranges and 4 lemons cost 43 zeds	–								✓					
Item 16 (M052090)		2011, 2015	MC	Set up system of equations	–								✓					

(continued)

Table 4.22 (continued)

Item	Grade level	Assessment year(s)	Item format	Item description	Figure	Mathematics errors and misunderstandings												
						M1	M2A	M2B	M3A	M3B	M4A	M4B	M5	M6	M7A	M7B	M8	M9
Performance objective 6: Given pairs of numbers in tables or ordered pairs, generate an algebraic equation of the relationship between the two variables																		
Item 17 (M032163)	Grade 8	2003, 2007	MC	Relation between x and y in a table	4.34									✓				
Item 18 (M022042)		1999	MC	Correct equation based on x/y table	–									✓				
Item 19 (M012046)		1995, 1999	MC	Equation from x/y table	4.35									✓				
Performance objective 7: Given pairs of numbers in tables or ordered pairs, generate a verbal description of the relationship																		
Item 20 (M012029)	Grade 8	1995, 1999, 2003	MC	Sets of ordered pairs of numbers	4.36										✓			
Item 21 (M041125)	Grade 4	2007, 2011, 2015	MC	Rule to get numbers in column B	4.37											✓		
Item 22 (M031227)		2003, 2007	CR	Sean's rule to transform numbers	4.38											✓		
Item 23 (M031251)		2003, 2007, 2011	MC	Ebru's rule to get number in box	–											✓		
Performance objective 8: Given a verbal description of a relationship between a set of numbers, generate pairs of whole numbers that follow that relationship (rule)																		
Item 24 (M041124)	Grade 4	2007, 2011, 2015	CR	Use the rule to complete the table	4.39												✓	
Item 25 (M061254)[a]		2015	CR	Complete the table for Mia's input-output game	–												✓	
Item 26 (M031242A)		2003, 2007	CR	Renting bikes/fill table	4.40												✓	

(continued)

Table 4.22 (continued)

Item	Grade level	Assessment year(s)	Item format	Item description	Figure	Mathematics errors and misunderstandings												
						M1	M2A	M2B	M3A	M3B	M4A	M4B	M5	M6	M7A	M7B	M8	M9
Performance objective 9: Apply algebraic thinking to solve simple real-life problems involving unknowns																		
Item 27 (M031247)	Grade 4	2003, 2007	CR	A man took his children to a fair	4.41													✓
Item 28 (M051006)		2011, 2015	CR	Cost of ice cream	4.42													✓

Notes See Table 4.21 for a full explanation of the codes relate to mathematics errors and misunderstandings. For each item, the relevant TIMSS item ID code as used in the TIMSS and TIMSS Advanced databases and released item sets is shown in brackets. For mathematics item 1, parts A and B were scored together as one item in the TIMSS data files. However, for the misconception analyses in this report, parts A and B are treated as two separate items

✓ Indicates the specific type of error or misunderstanding is measured by the item; *MC* multiple choice; *CR* constructed response

[a]Item is secured

4.3.1 Student Performance on TIMSS and TIMSS Advanced Items Related to Linear Equations

The performance of students on the set of linear equation items across grade levels covered a broad range both within and across countries (Fig. 4.25, 4.26, and 4.27), with some very difficult items (<20% correct) and some easier items ($\geq 60\%$

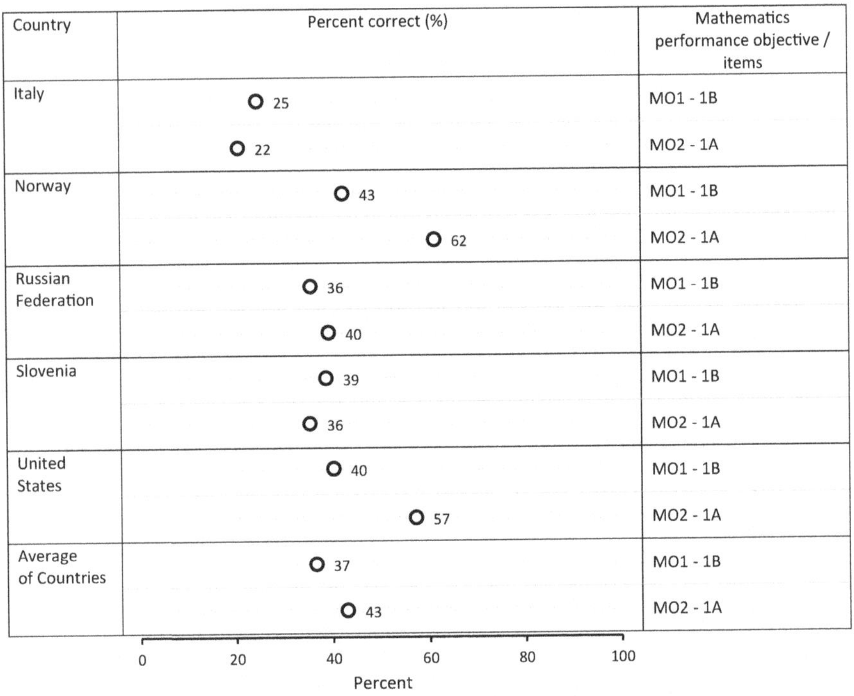

Fig. 4.25 Student performance on TIMSS Advanced mathematics items, by country and performance objective, 2015. *Notes* Percent correct is the percentage of students receiving credit on each item. For MC and short CR items (each worth one score point), this reflects the percentage of students who provided a correct answer. For extended CR items, this reflects the weighted percentage of students receiving full credit (2 points) or partial credit (1 point). The percentages are for the most recent cycle each item was administered. Data for item 1A and 1B are from 2015; Item 1 (parts A and B) was scored using an overall scoring guide (shown in Fig. 4.28). The percent correct shown for 1A reflects all students who answered part A correctly (codes 20 and 10 combined). The percent correct shown for 1B reflects all students who answered part B correctly (codes 20 and 11 combined). Mathematics performance objectives (MO): MO1 = interpret the solution to a system of linear equations to answer a question or solve a problem in real life contexts, MO2 = solve systems of linear equations in two variables

correct).[20] Average performance across the five countries for the TIMSS Advanced items (Fig. 4.25) ranged from 37 to 43% correct. In comparison, the average item performance on grade eight items (Fig. 4.26) ranged from 13 to 63% correct and on grade four items (Fig. 4.27) ranged from 17 to 62% correct. Some notable differences in performance were observed across the five countries.

For TIMSS Advanced (Fig. 4.25), the widest-ranging item performance was in Norway (from 43 to 62% correct) and the United States (from 40 to 57% correct). In contrast, item-level performance ranged from 36 to 40% correct in the Russian Federation, from 22 to 25% correct in Italy, and from 36 to 39% correct in Slovenia. In three of the five countries, students found item 1B more difficult than item 1A. In contrast, in Italy and Slovenia, 3% more students found item 1A more difficult than item 1B.

At grade eight (Fig. 4.26), a broad range of item performance was found in all five countries, with the lowest range (43 percentage points) in Norway and the highest range (61 percentage points) in the Russian Federation and the United States. The three most difficult items in all countries were CR items. Two of them, items 2 and 3, are from performance objective 2 ("solve system of linear equations in two variables") and one, item 6, is from performance objective 3 ("interpret the meanings of the slope and y-intercept in linear equations and graphs"). For both items 2 and 3, item performance was lowest in Norway (4% and 1% correct, respectively) and highest in the Russian Federation (36% correct on both items). On item 6, performance was lowest in Italy (2% correct) and highest in the United States (31% correct). In comparison, the two easiest items (item 12 and item 13) were both MC items. Item 12 is from performance objective 4 ("relate algebraic equations to their graphical representations (and vice versa)") and item 13 is from objective 5 ("write equations to represent situations"). For items 12 and 13, performance was lowest in Norway (44% and 41% correct, respectively) and highest in the United States (78% and 75% correct, respectively).

There was also a broad range of item performance found in each country at grade four (Fig. 4.27). The smallest range in percent correct across grade four items was in the United States (23–66%) and the largest was in the Russian Federation (22–81%). Interestingly, at grade four, the three most difficult items across countries (items 22, 27, and 28) as well as the two of the easiest items (items 23 and 26) were CR items. In comparison, at grade eight, the easiest items were all MC items. One of the most difficult items at grade four, item 22, is from performance objective 7 ("given pairs of numbers in tables or ordered pairs, generate a verbal description of the relationship"). The performance on this item varied from 8% correct in Slovenia to 23% correct in the Russian Federation and the United States. The other two items are from performance objective 9 ("apply algebraic thinking to solve simple real-life problems involving unknowns"). The items that were easy in general for all countries were from performance objective 8 ("given a verbal description of a relationship between a set of numbers, generate pairs of whole numbers that follow that relationship/rule").

---

[20]The data presented in Fig. 4.25, 4.26, and 4.27 reflect the most recent assessment in which each item was administered from 1995 to 2015. See Table 4.22 for the most recent assessment for each item. Changes in performance between assessment cycles for trend items are reported in Sect. 4.2.5.

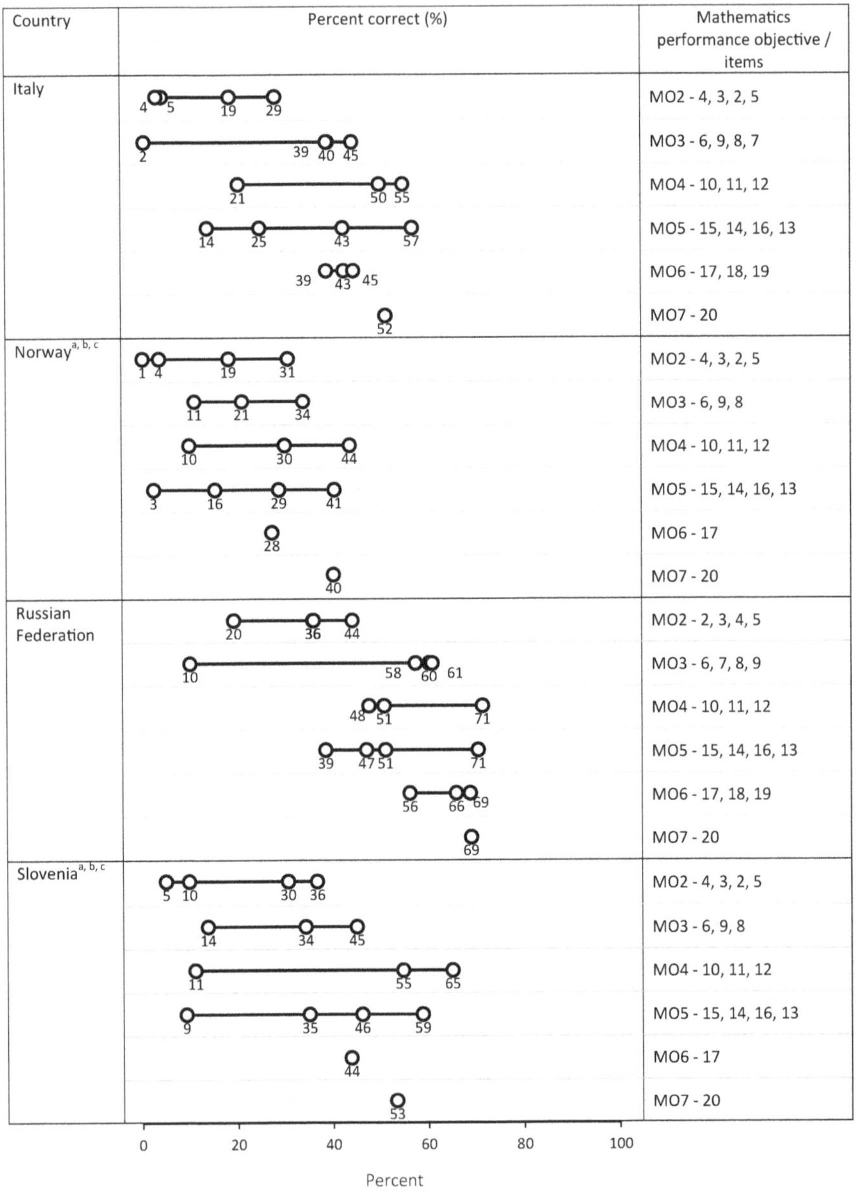

Country
Percent correct (%)
Mathematics performance objective / items
Italy
4 5 19 29
MO2 - 4, 3, 2, 5
2 39 40 45
MO3 - 6, 9, 8, 7
21 50 55
MO4 - 10, 11, 12
14 25 43 57
MO5 - 15, 14, 16, 13
39 43 45
MO6 - 17, 18, 19
52
MO7 - 20
Norway[a, b, c]
1 4 19 31
MO2 - 4, 3, 2, 5
11 21 34
MO3 - 6, 9, 8
10 30 44
MO4 - 10, 11, 12
3 16 29 41
MO5 - 15, 14, 16, 13
28
MO6 - 17
40
MO7 - 20
Russian Federation
20 36 44
MO2 - 2, 3, 4, 5
10 58 60 61
MO3 - 6, 7, 8, 9
48 51 71
MO4 - 10, 11, 12
39 47 51 71
MO5 - 15, 14, 16, 13
56 66 69
MO6 - 17, 18, 19
69
MO7 - 20
Slovenia[a, b, c]
5 10 30 36
MO2 - 4, 3, 2, 5
14 34 45
MO3 - 6, 9, 8
11 55 65
MO4 - 10, 11, 12
9 35 46 59
MO5 - 15, 14, 16, 13
44
MO6 - 17
53
MO7 - 20
0 20 40 60 80 100
Percent

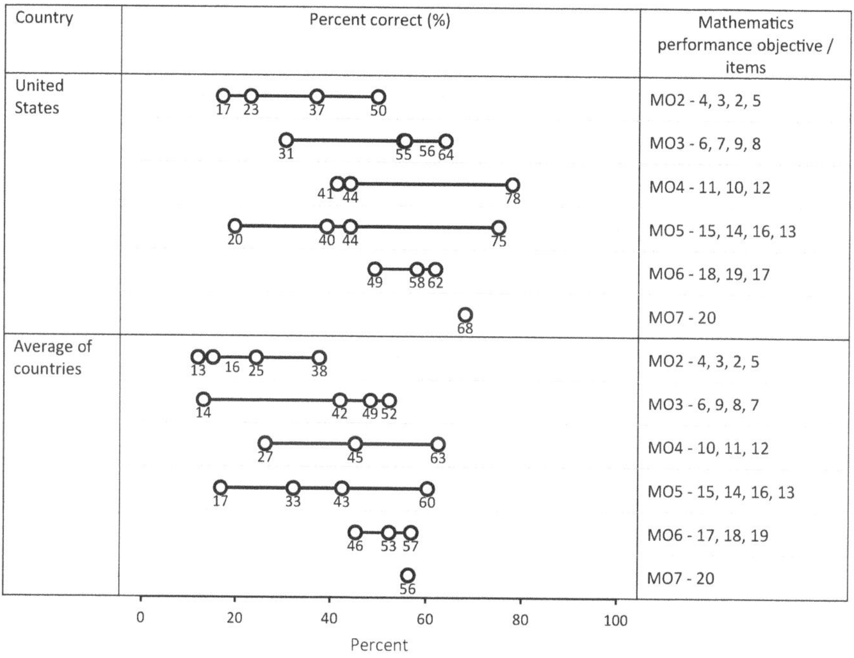

Fig. 4.26 (continued)

◂**Fig. 4.26** Student performance on TIMSS grade eight mathematics items, by country and performance objective, 1999, 2003, 2007, and 2015. *Notes* Percent correct is the percentage of students receiving credit on each item. For MC and short CR items (each worth one score point), this reflects the percentage of students who provided a correct answer. For extended CR items, this reflects the weighted percentage of students receiving full credit (2 points) or partial credit (1 point). The percentages are for the most recent cycle each item was administered. Data for items 3, 4, 6, 8–14, and 16 are from 2015; data for items 2 and 17 are from 2007; data for items 5, 15, and 20 are from 2003; and data for items 7, 18, and 19 are from 1999. Mathematics performance objective (MO): MO2 = solve systems of linear equations in two variables, MO3 = interpret the meanings of slope and y-intercept in linear equations or graphs, MO4 = relate algebraic equations to their graphical representations (and vice versa), MO5 = write equations to represent situations, MO6 = given pairs of numbers in tables or ordered pairs, generate an algebraic equation of the relationship between the two variables. [a]Data not available for item 7 (see Appendix for country-specific notes). [b]Data not available for item 18 (see Appendix for country-specific notes). [c]Data not available for item 19 (see Appendix for country-specific notes)

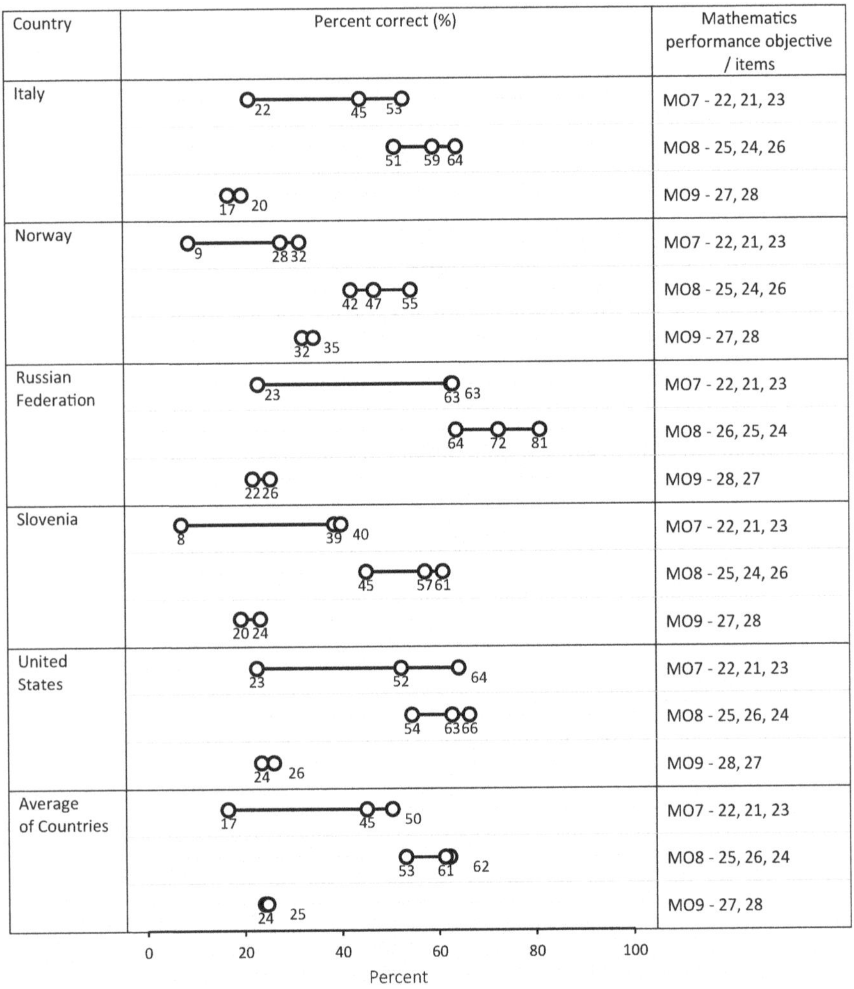

Fig. 4.27 Student performance on TIMSS grade four mathematics items, by country and performance objective, 2007, 2011, and 2015. *Notes* Percent correct is the percentage of students receiving credit on each item. For MC and short CR items (each worth one score point), this reflects the percentage of students who provided a correct answer. For extended CR items, this reflects the weighted percentage of students receiving full credit (2 points) or partial credit (1 point). The percentages are for the most recent cycle each item was administered. Data for items 21, 24, 25, and 28 are from 2015; data for item 23 are from 2011; and data for items 22, 26, and 27 are from 2007. Mathematics performance objective (MO): MO7 = given pairs of numbers in tables or ordered pairs, generate a verbal description of the relationship, MO8 = given a verbal description of a relationship between a set of numbers, generate pairs of whole numbers that follow that relationship (rule), MO9 = apply algebraic thinking to solve simple real-life problems involving unknowns

4.3.2 Common Types of Errors and Misunderstandings Related to Linear Equations Across Countries

By the time students reach upper-secondary school, they are expected to be well versed with linear equations/relationships. They should be able to understand solving equations as a process of reasoning and explain their reasoning. The first example item (item 1: Fig. 4.28 and Table 4.23) shows that many students still have difficulty in meeting this expectation. A fully correct score (code 20 in the scoring guide) requires students to correctly answer both parts A and B. A correct response to part A must provide the correct answer with adequate work shown (algebraically or graphically). The algebraic solution includes writing equations for the two different car rental plans (X and Y) and then solving the simultaneous equations to find the point at which they intersect (3000 km) or have the same cost. In part B, a correct response requires students to understand and explain that if the same increase in initial cost is applied to both plans with no other change, the difference between the two y-intercepts remains the same. Therefore, the distance on the x-axis at which the two equations intersect will not change. A common misunderstanding is that the students do not understand the solution well enough to go beyond the set procedure and explain the solution with a change in the initial cost of the rental plan.

Students received partial credit for providing a correct response to either part A (code 10) or part B (code 11). An incorrect response to part A (code 11, 79, or 99) demonstrates that students cannot evaluate the context, write equations, or apply the correct procedures to solve a system of equations. On average, 58% of students demonstrated this error across the five countries. The percentage of students making this error varied considerably across the five countries, from 79% of students in Italy, to 38% of students in Norway and 43% of students in the United States.

An incorrect response to part B (code 10, 79, or 99) demonstrates that students did not understand the system of equations well enough to explain the impact of the y-intercept change on both equations. That is, they did not demonstrate a deeper understanding beyond applying procedures to solve a pair of linear equations. On average, 64% of students demonstrated this error across the five countries. Italy had the highest percentage of students (76%) demonstrating this misunderstanding, and Norway had the lowest percentage (57%).

By grade eight, students are expected to be able to solve a given system of linear equations in two variables in context as well as not in context. However, many grade eight students made errors in doing so on the TIMSS assessments. This is shown in the first grade eight example item (item 2: Fig. 4.29 and Table 4.24). Similar to the first TIMSS Advanced item (Part A only), this item asks students to write simultaneous linear equations to represent the given situation, and then solve them to get the cost of one pen and two pencils. The important thing is that they need to show their work in order to receive a correct score. On average across the five countries, 25% of students (codes 10 and 11 in the scoring guide) were able to correctly solve this problem and show their work in support of their answers. An

Two different plans for renting a car are given in the table below.

Rental Plan	Initial Cost	Cost per kilometer
X	100 zeds	.07 zeds
Y	250 zeds	.02 zeds

A. After how many kilometers does Plan Y become the cheaper plan?

Show your work.

B. If an extra insurance charge of 100 zeds is charged by both plans, does this change the number of kilometers when Plan Y becomes cheaper?

Explain your answer.

Item information

Item ID:
MA33240 (A & B)

Year(s) administered:
2015

Performance objective:
Part A. Solve systems of linear equations in two variables.

Part B. Interpret the solution to a system of linear equations to answer a question or solve a problem in real life contexts

Scoring guide

Correct response	
20	Completes both part A and part B correctly. A. 3000 or equivalent (e.g., 150/0.05) kilometers with adequate work shown (Accept $x > 3000$ and $x = 3001$) B. No and explains that the number of kilometers does not change when Y becomes the cheaper plan
Partial Response	
10	Part A correct only
11	Part B correct only
Incorrect response	
79	Incorrect for both part A and part B (including crossed out, erased, stray marks, illegible, or off task)
Non response	
99	Blank

Fig. 4.28 TIMSS Advanced mathematics item 1, 2015. *Source* TIMSS Advanced 2015 Assessment. Copyright © 2017 International Association for the Evaluation of Educational Achievement (IEA). TIMSS & PIRLS International Study Center, Lynch School of Education, Boston College

Table 4.23 Student performance data for mathematics item 1 (MA33240A and MA33240B), 2015

Country	Percentage of students (%)							
	Correct (weighted)	Correct Part A (20, 10)	Correct Part B (20, 11)	20	10	11	79	99
Italy	23	↓ 22	25	18	4	7	23	49
Norway	52	↑ 62	43	39	22	3	19	16
Russian Federation	38	40	36	30	10	6	22	32
Slovenia	37	36	39	28	8	11	33	20
United States	49	↑ 57	40	35	22	5	31	7
Average of countries	40	43	37	30	13	7	26	25

Notes Correct (weighted) reflects the weighted percentage of students receiving full credit (code 20) or partial credit (code 10 or 11). Detail may not sum to totals due to rounding. Although the table displays rounded data, the calculations of the combined correct responses and weighted percent correct are based on unrounded data

↑ Significantly higher than the average of countries
↓ Significantly lower than the average of countries

Joe knows that a pen costs 1 zed more than a pencil.
His friend bought 2 pens and 3 pencils for 17 zeds.
How many zeds will Joe need to buy 1 pen and 2 pencils?

Show your work.

Item information

Item ID:
M042263

Year(s) administered:
2007

Performance objective:
Solve systems of linear equations in two variables

Scoring guide

Correct response	
10	10 zeds and equation(s) shown. Equations should involve the use of letter(s) as variable(s), e.g., $2y + 3x = 17$.
11	10 zeds and other work shown, e.g., pen = pencil + 1
Incorrect response	
70	10 zeds, no work shown
79	Incorrect (including crossed out, erased, stray marks, illegible, or off task)
Non response	
99	Blank

Fig. 4.29 TIMSS grade eight mathematics item 2, 2007. *Source* TIMSS 2007 Assessment. Copyright © 2009 International Association for the Evaluation of Educational Achievement (IEA). TIMSS & PIRLS International Study Center, Lynch School of Education, Boston College

Table 4.24 Student performance data for mathematics item 2 (M042263), 2007

Country	Percentage of students (%)					
	Correct (10-11)	10	11	70	79	99
Italy	19 ▼	3	17	3	35	43
Norway	19 ▼	2	17	5	44	32
Russian Federation	20 ▼	18	2	5	35	40
Slovenia	30 ▲	2	28	2	50	18
United States	37 ▲	5	32	2	51	10
Average of countries	25	6	19	3	43	29

Notes Detail may not sum to totals due to rounding. Although the table displays rounded data, the calculations of the combined correct responses are based on unrounded data
▲ Significantly higher than the average of countries
▼ Significantly lower than the average of countries

additional 3% of students obtained the correct answer but did not show their work (code 70). On average, 72% of students were not able to correctly solve the problem (code 79 and 99 in the scoring guide). The percentage of students not able to complete this task was high across all five countries, with the United States having the lowest percentage (61%) and Italy having the highest (78%).

In the second grade eight example (item 3: Fig. 4.30 and Table 4.25), students were expected to solve the system of linear equations provided. Very few students across the five countries could apply the procedure correctly in order to obtain the solution. A high percentage of students either made a mistake (code 79 in the scoring guide) or did not attempt the problem at all (code 99). On average, 85% of students were not able to correctly solve the given system of equations across the five countries; the highest percentage of students unable to solve the problems was in Norway (96%), and the lowest was in the Russian Federation (64%).

Find the values of x and y such that both equations are true.

$$3x + y = 13$$
$$5x - y = 27$$

$x =$ __________

$y =$ __________

Item information
Item ID:
M062237

Year(s) administered:
2015

Performance objective:
Solve systems of linear equations in two variables

Scoring guide

Correct response	
10	$x = 5$ $y = -2$
Incorrect response	
79	Incorrect (including crossed out, erased, stray marks, illegible, or off task)
Non response	
99	Blank

Fig. 4.30 TIMSS grade eight mathematics item 3, 2015. *Source* TIMSS 2015 Assessment. Copyright © 2017 International Association for the Evaluation of Educational Achievement (IEA). TIMSS & PIRLS International Study Center, Lynch School of Education, Boston College

Table 4.25 Student performance data for mathematics item 3 (M062237), 2015

Country	Percentage of students (%)			
	Correct (10)	10	79	99
Italy	5 ↓	5	46	49
Norway	4 ↓	4	53	43
Russian Federation	36 ↑	36	27	37
Slovenia	10 ↓	10	61	29
United States	23 ↑	23	66	10
Average of countries	16	16	51	34

Notes Detail may not sum to totals due to rounding
↑ Significantly higher than the average of countries
↓ Significantly lower than the average of countries

Another performance objective is that, by the end of grade eight, students should be able to identify and interpret the slope and the intercepts in linear equations shown algebraically and graphically. Item 6 (not shown as it is a secured item) includes two lines on a graph and their equations. Students were expected to determine which line had a greater slope by relating the steepness of the line with the slope of the linear equation. On average across the five countries, nearly 86% of the students failed to correctly identify which slope was larger. In all countries except the United States (69%), more than 85% of the students were not able to correctly relate that the steeper line has the larger slope.

In the next grade eight example (item 9: Fig. 4.31 and Table 4.26), students were expected to identify and select the correct equation of a line based on a verbal description of the conditions given. On average across the five countries, 57% of the students got this item correct, with the lowest percentage correct being in Norway (52%) and the highest in the United States (64%). Two of the distractors (options A and C) were not the equation of a line, and hence could be eliminated. The other distractor (option B) had the intercept and the slope swapped in

Janet described the graph of a function:

- The graph is a straight line.
- The graph intercepts the y-axis at 3.

Which could be the function of Janet's graph?

Ⓐ $y = x^2 + 3$

Ⓑ $y = 3x + 1$

Ⓒ $y = 3x^2 - 1$

Ⓓ $y = x + 3$

Item information

Item ID:
M062074

Year(s) administered:
2015

Performance objective:
Interpret the meanings of slope and y-intercept in linear equations or graphs

Correct answer:
D

Fig. 4.31 TIMSS grade eight mathematics item 9, 2015. *Source* TIMSS 2015 Assessment. Copyright © 2017 International Association for the Evaluation of Educational Achievement (IEA). TIMSS & PIRLS International Study Center, Lynch School of Education, Boston College

Table 4.26 Student performance data for mathematics item 9 (M062074), 2015

Country	Percentage of students (%)						
	Correct (D)		A	B	C	D	Omitted
Italy	39		7	27	14	39	12
Norway	21	↓	9	29	22	21	19
Russian Federation	61	↑	13	11	9	61	6
Slovenia	34	↓	10	20	20	34	15
United States	56	↑	9	22	9	56	5
Average of countries	42		9	22	15	42	11

Notes Detail may not sum to totals due to rounding
↑ Significantly higher than the average of countries
↓ Significantly lower than the average of countries

comparison to the correct equation (option D). On average, 22% of students chose option B; the Russian Federation had the lowest percentage of students choosing option B (11%), while the highest percentage of students choosing this option was in Norway (29%).

Another performance objective is that by the end of grade eight, students should be able to translate between algebraic and graphical representations. In item 10 (not shown as it is a secured item), students need to identify the graph of a given equation. On average, across the five countries, only 23% of students correctly identified the graph of the equation given in the item; students in Norway were least likely to get this correct (10%), while the highest percentage of students getting this correct was in the Russian Federation (48%).

Another grade eight example (item 11: Fig. 4.32 and Table 4.27) required students to identify the equation of a line whose graph was given. On average across the five countries, 45% of students got this item correct; Norway had the lowest

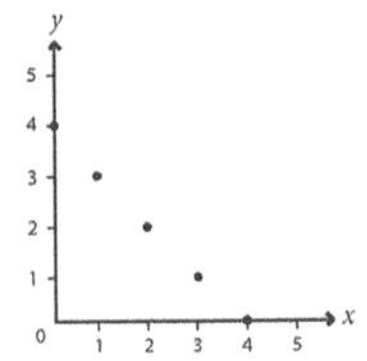

x and *y* are whole numbers.

What is the relationship between *x* and *y* for the points shown on the graph?

Ⓐ $y = x - 4$

Ⓑ $x = y - 4$

Ⓒ $x + 4y = 4$

Ⓓ $x + y = 4$

Item information

Item ID:
M042112

Year(s) administered:
2015, 2011, 2007

Performance objective:
Relate algebraic equations to their graphical representations (and vice versa)

Correct answer:
D

Fig. 4.32 TIMSS grade eight mathematics item 11, 2015. *Source* TIMSS 2015 Assessment. Copyright © 2017 International Association for the Evaluation of Educational Achievement (IEA). TIMSS & PIRLS International Study Center, Lynch School of Education, Boston College

Table 4.27 Student performance data for mathematics item 11 (M042112), 2015

Country	Percentage of students (%)					
	Correct (D)	A	B	C	D	Omitted
Italy	50 ↑	10	15	15	50	10
Norway	30 ↓	13	22	24	30	11
Russian Federation	51 ↑	13	10	17	51	9
Slovenia	55 ↑	10	16	12	55	7
United States	41 ↓	26	14	16	41	2
Average of countries	45	14	15	17	45	8

Notes Detail may not sum to totals due to rounding
↑ Significantly higher than the average of countries
↓ Significantly lower than the average of countries

percentage of correct responses (30%) and Slovenia the highest percentage (55%). In all countries, about half of the students were not able to correctly identify the equation of line for the given graph.

In the next grade eight example (item 13: Fig. 4.33 and Table 4.28), students are expected to identify the correct equation of a line from the verbal description of the given situation. More than 60% of students, on average across the five countries, were able to identify the correct equation of the line (option B). Performance on this item varied considerably across countries; Norway had the lowest percentage correct (41%) and the United States the highest percentage correct (75%). The most common incorrect response in all countries was option D, which was selected by 27% of students on average.

A class visits a museum. Lunch for the whole class costs B zeds. It costs 4 zeds per student to pay for entry to the museum. There are p students in the class.

The total cost for the trip is K zeds. What is the formula for K?

Ⓐ $K = B + 4$

Ⓑ $K = B + 4p$

Ⓒ $K = B + p$

Ⓓ $K = (B + p) \times 4$

Item information

Item ID:
M042202

Year(s) administered:
2015, 2011, 2007

Performance objective:
Write equations to represent situations

Correct answer:
B

Fig. 4.33 TIMSS grade eight mathematics item 13, 2015. *Source* TIMSS 2015 Assessment. Copyright © 2017 International Association for the Evaluation of Educational Achievement (IEA). TIMSS & PIRLS International Study Center, Lynch School of Education, Boston College

Table 4.28 Student performance data for mathematics item 13 (M042202), 2015

Country	Percentage of students (%)						
	Correct (B)		A	B	C	D	Omitted
Italy	57		3	57	5	33	2
Norway	41	↓	8	41	11	33	8
Russian Federation	71	↑	2	71	1	24	2
Slovenia	59		4	59	5	32	1
United States	75	↑	5	75	5	14	1
Average of countries	60		4	60	5	27	3

Notes Detail may not sum to totals due to rounding
↑ Significantly higher than the average of countries
↓ Significantly lower than the average of countries

The next two grade eight example items required students to identify the correct equation of a line from the relationship given in a table. In item 17 (Fig. 4.34 and Table 4.29), all the equations in the response options are for a straight line. On average across the five countries, 46% of the students got this item correct (option C), but performance varied considerably. The percent correct in Norway (28%) and Italy (39%) was significantly lower than the average across the five countries, while that for the Russian Federation (56%) and the United States (62%) was higher than average. Item 19 (Fig. 4.35 and Table 4.30)[21] also required students to identify the correct equation of a line from the relationship given in a table. However, in this

The table below shows a relation between x and y.

x	1	2	3	4	5
y	1	3	5	7	9

Which of the following equations expresses this relation?

Ⓐ $y = x + 4$

Ⓑ $y = x + 1$

Ⓒ $y = 2x - 1$

Ⓓ $y = 3x - 2$

Item information
Item ID:
M032163

Year(s) administered:
2007, 2003

Performance objective:
Given pairs of numbers in tables or ordered pairs, generate an algebraic equation of the relationship between the two variables

Correct answer:
C

Fig. 4.34 TIMSS grade eight mathematics item 17, 2007. *Source* TIMSS 2007 Assessment. Copyright © 2009 International Association for the Evaluation of Educational Achievement (IEA). TIMSS & PIRLS International Study Center, Lynch School of Education, Boston College

Table 4.29 Student performance data for mathematics item 17 (M032163), 2007

Country	Percentage of students (%)					
	Correct (C)	A	B	C	D	Omitted
Italy	39 ⇓	16	23	39	11	10
Norway	28 ⇓	22	29	28	9	12
Russian Federation	56 ⇑	14	19	56	5	6
Slovenia	44	13	28	44	10	6
United States	62 ⇑	12	17	62	8	1
Average of countries	46	15	23	46	9	7

Notes Detail may not sum to totals due to rounding
⇑ Significantly higher than the average of countries
⇓ Significantly lower than the average of countries

[21] Slovenia and Norway did not participate in TIMSS 1999 assessment. Hence, average performance for this item is based on data from three countries (Italy, the Russian Federation, and the United States).

The table shows a relation between x and y.

x	2	3	4	5
y	7	10	13	16

Which of these equations expresses this relation?

A. $y = x + 5$

B. $y = x - 5$

C. $y = \frac{1}{3}(x - 1)$

D. $y = 3x + 1$

Item information
Item ID:
M012046

Year(s) administered:
1999, 1995

Performance objective:
Given pairs of numbers in tables or ordered pairs, generate an algebraic equation of the relationship between the two variables

Correct answer:
D

Fig. 4.35 TIMSS grade eight mathematics item 19, 1999. *Source* TIMSS 1999 Assessment. Copyright © 2001 International Association for the Evaluation of Educational Achievement (IEA). TIMSS & PIRLS International Study Center, Lynch School of Education, Boston College

Table 4.30 Student performance data for mathematics item 19 (M012046), 1999

Country	Percentage of students (%)						
	Correct (D)		A	B	C	D	Invalid
Italy	45	▼	20	15	20	45	0
Norway	—		—	—	—	—	—
Russian Federation	69	▲	20	5	6	69	1
Slovenia	—		—	—	—	—	—
United States	58		20	10	11	58	0
Average of countries	57		20	10	13	57	0

Notes Detail may not sum to totals due to rounding
▲ Significantly higher than the average of countries
▼ Significantly lower than the average of countries
— Data not available for item 19 (see Appendix for country-specific notes)

case, two response options (options B and C) are not equations for a single straight line. Hence, the choice is between the other two options.

By the end of elementary school, students are expected to be able to express the relationship between ordered pairs or between two numbers (input/output numbers). The last grade eight example (item 20: Fig. 4.36 and Table 4.31) asked students to identify the correct relationship between the set of given ordered pairs. On average, 56% of the students answered correctly. Performance in Norway (40% correct) and Italy (52% correct) was lower than the international average, while performance in the United States (68% correct) and the Russian Federation (69% correct) was higher than average.

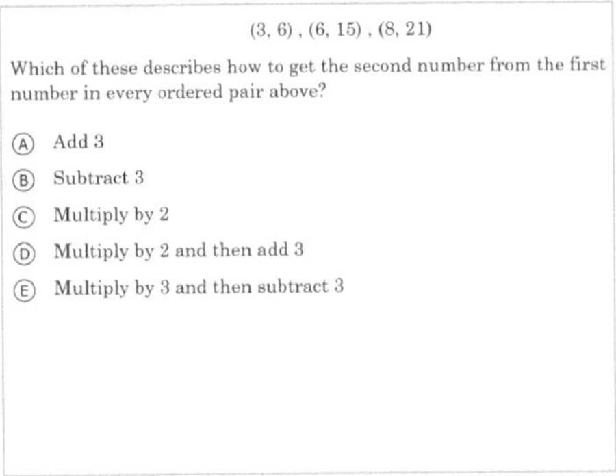
(3, 6) , (6, 15) , (8, 21)

Which of these describes how to get the second number from the first number in every ordered pair above?

Ⓐ Add 3
Ⓑ Subtract 3
Ⓒ Multiply by 2
Ⓓ Multiply by 2 and then add 3
Ⓔ Multiply by 3 and then subtract 3

Item information
Item ID:
M012029

Year(s) administered:
2003, 1999, 1995

Performance objective:
Given pairs of numbers in tables or ordered pairs, generate a verbal description of the relationship

Correct answer:
E

Fig. 4.36 TIMSS grade eight mathematics item 20, 2003. *Source* TIMSS 2003 Assessment. Copyright © 2005 International Association for the Evaluation of Educational Achievement (IEA). TIMSS & PIRLS International Study Center, Lynch School of Education, Boston College

Table 4.31 Student performance data for mathematics item 20 (M012029), 2003

Country	Percentage of students (%)								
	Correct (E)		A	B	C	D	E	Invalid	Omitted
Italy	52	∪	4	4	9	19	52	3	9
Norway	40	∪	8	4	18	21	40	0	8
Russian Federation	69	∩	4	3	5	10	69	3	6
Slovenia	53		5	5	8	19	53	3	7
United States	68	∩	4	4	7	17	68	0	1
Average of countries	56		5	4	9	17	56	2	6

Notes Detail may not sum to totals due to rounding
∩ Significantly higher than the average of countries
∪ Significantly lower than the average of countries

At grade four, there are two items (21 and 22) assessing the same performance objective as the previous grade eight item 20. Item 21 (Fig. 4.37 and Table 4.32) is a MC item that required students to correctly identify the verbal description of a rule to obtain a number in column B from a number in column A. On average across the five countries, 45% of students were able to identify the correct description of the rule (option A) from the four choices given. There was a considerable range in performance across countries on this item; in Norway 28% of students responded correctly, while at the other end, in the Russian Federation, 63% of students responded correctly. In item 22 (Fig. 4.38 and Table 4.33), which is a CR item, students were provided with a set of four paired numbers and asked to write a verbal description of the rule applied to the first number in each pair in order to obtain the second number. Student performance on this item was extremely low and there was a high percentage of blank responses (code 99 in the scoring guide). On average across the five countries, only 17% of students were able to provide the

Look at this table of numbers.

Column A	Column B
1	2
2	5
3	10
4	17

Which rule gives the number in Column B?

Ⓐ Multiply the number in Column A by itself, then add 1.
Ⓑ Multiply the number in Column A by 3, then subtract 1.
Ⓒ Multiply the number in Column A by itself, then subtract 1.
Ⓓ Multiply the number in Column A by 2.

Item information

Item ID:
M041125

Year(s) administered:
2015, 2011, 2007

Performance objective:
Given pairs of numbers in tables or ordered pairs, generate a verbal description of the relationship

Correct answer:
A

Fig. 4.37 TIMSS grade four mathematics item 21, 2015. *Source* TIMSS 2015 Assessment. Copyright © 2017 International Association for the Evaluation of Educational Achievement (IEA). TIMSS & PIRLS International Study Center, Lynch School of Education, Boston College

Table 4.32 Student performance data for mathematics item 21 (M041125), 2015

Country	Percentage of students (%)					
	Correct (A)	A	B	C	D	Omitted
Italy	45	45	23	12	12	9
Norway	28 ↓	28	26	17	16	12
Russian Federation	63 ↑	63	18	5	10	4
Slovenia	39 ↓	39	21	10	18	13
United States	52 ↑	52	24	9	13	2
Average of countries	45	45	22	11	14	8

Notes Detail may not sum to totals due to rounding
↑ Significantly higher than the average of countries
↓ Significantly lower than the average of countries

correct response. Performance in Italy, the Russian Federation, and the United States was very similar (22–23% correct), and the performance of students in Norway and Slovenia (8–9% correct) was also very similar.

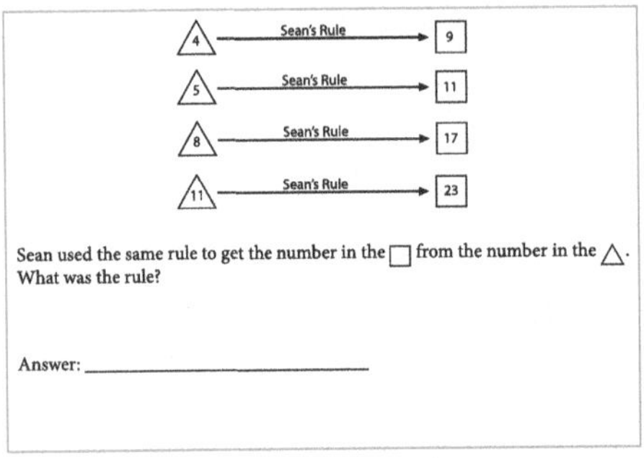

Item information

Item ID:
M031227

Year(s) administered:
2007, 2003

Performance objective:
Given pairs of numbers in tables or ordered pairs, generate a verbal description of the relationship

Scoring guide

Correct response	
10	Double the number in the triangle and add 1 (e.g., double and add 1; multiply by 2 and add 1)
19	Other correct, including adding the next highest number to the given number in the triangle (e.g., 4 + 5 = 9)
Incorrect response	
79	Incorrect (including crossed out, erased, stray marks, illegible, or off task)
Non response	
99	Blank

Fig. 4.38 TIMSS grade four mathematics item 22, 2007. *Source* TIMSS 2007 Assessment. Copyright © 2009 International Association for the Evaluation of Educational Achievement (IEA). TIMSS & PIRLS International Study Center, Lynch School of Education, Boston College

Table 4.33 Student performance data for mathematics item 22 (M031227), 2007

Country	Percentage of students (%)				
	Correct (10-19)	10	19	79	99
Italy	22 ↑	19	2	47	32
Norway	9 ↓	7	3	55	36
Russian Federation	23 ↑	20	3	54	22
Slovenia	8 ↓	6	2	63	29
United States	23 ↑	16	6	70	7
Average of countries	17	14	3	58	25

Notes Detail may not sum to totals due to rounding. Although the table displays rounded data, the calculations of the combined correct responses are based on unrounded data

↑ Significantly higher than the average of countries

↓ Significantly lower than the average of countries

A related objective at grade four is that students should be able to use a given rule to obtain the output numbers from the input numbers (item 24: Fig. 4.39 and Table 4.34). On average across the five countries, 62% of students were able to provide both correct entries in column B of the table, and another 6% were able to provide one correct entry. An additional 4% of students were able to apply the partial or incomplete rule (i.e., they multiplied the number in column A by 4, but then forgot to add 1 to the result). Student performance on this item varied

Rule: To find the number in Column B, multiply the number in Column A by 4, then add 1.

Use this rule to fill in the table below.

Column A	Column B
2	
5	

Item information
Item ID:
M041124

Year(s) administered:
2015, 2011, 2007

Performance objective:
Given a verbal description of a relationship between a set of numbers, generate pairs of whole numbers that follow that relationship (rule)

Scoring guide

Correct response	
10	Both entries correct: 9 21
Incorrect response	
70	1 only correct
71	8 20
79	Other incorrect (including crossed out, erased, stray marks, illegible, or off task)
Non response	
99	Blank

Fig. 4.39 TIMSS grade four mathematics item 24, 2015. *Source* TIMSS 2015 Assessment. Copyright © 2017 International Association for the Evaluation of Educational Achievement (IEA). TIMSS & PIRLS International Study Center, Lynch School of Education, Boston College

Table 4.34 Student performance data for mathematics item 24 (M041124), 2015

Country	Percentage of students (%)					
	Correct (10)	10	70	71	79	99
Italy	59	59	5	8	22	6
Norway	47 ↓	47	9	4	33	7
Russian Federation	81 ↑	81	6	3	9	2
Slovenia	57 ↓	57	7	4	24	8
United States	66 ↑	66	6	4	21	2
Average of countries	62	62	6	4	22	5

Notes Detail may not sum to totals due to rounding
↑ Significantly higher than the average of countries
↓ Significantly lower than the average of countries

considerably across countries, covering a range of almost 34 percentage points. Student performance in Norway (47% correct) and Slovenia (57% correct) was below the international average, while that of the United States (66% correct) and the Russian Federation (81% correct) was above the international average.

Another example (item 26: Fig. 4.40 and Table 4.35) is from the same performance objective. This problem was placed in a real-life context, and students were expected to complete two table entries in order to receive credit for this item. On average across the five countries, 61% of students were able to complete both tables correctly (code 10 in the scoring guide). An additional 7% of students completed one of the tables correctly (codes 70 and 71). For four countries (Italy, the Russian Federation, Slovenia, and the United States), performance on this item was clustered between 61 and 64% correct. In contrast, in Norway, only 55% of students obtained the correct answer.

Posters for two sports clubs that rent bikes are shown below.

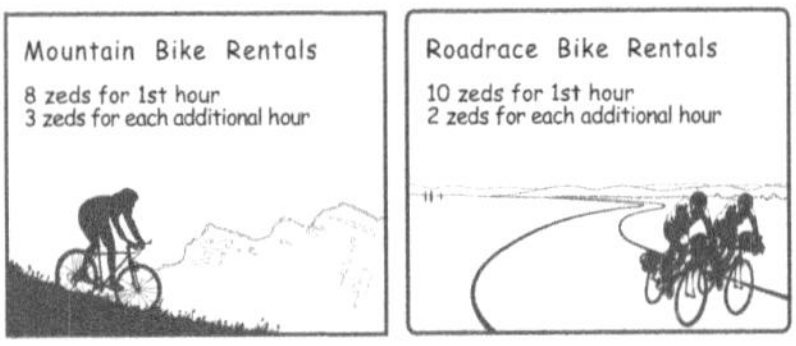

A. Use the information in the posters to complete the tables.

Mountain Bike Rentals	
Hours	Cost (zeds)
1	8
2	11
3	
4	
5	
6	

Roadrace Bike Rentals	
Hours	Cost (zeds)
1	10
2	12
3	
4	
5	
6	

Item information

Item ID:
M031242A

Year(s) administered:
2007, 2003

Performance objective:
Given a verbal description of a relationship between a set of numbers, generate pairs of whole numbers that follow that relationship (rule)

Scoring guide

Correct response					
10	Table completed correctly to 6 hours:	3 hours	14 zeds	3 hours	14 zeds
		4	17	4	6
		5	20	5	18
		6	23	6	20
Incorrect response					
70	One or more entries for Mountain Club incorrect; Roadrace Club entries all correct				
71	One or more entries for Roadrace Club incorrect; Mountain Club entries all correct				
79	Other incorrect (including crossed out, erased, stray marks, illegible, or off task)				
Non response					
99	Blank				

Fig. 4.40 TIMSS grade four mathematics item 26, 2007. *Source* TIMSS 2007 Assessment. Copyright © 2009 International Association for the Evaluation of Educational Achievement (IEA). TIMSS & PIRLS International Study Center, Lynch School of Education, Boston College

Table 4.35 Student performance data for mathematics item 26 (M031242A), 2007

Country	Percentage of students (%)					
	Correct (10)	10	70	71	79	99
Italy	64	64	5	2	14	15
Norway	55 ↓	55	7	3	21	15
Russian Federation	64	64	4	2	9	22
Slovenia	61	61	6	3	18	11
United States	63	63	5	2	11	19
Average of countries	61	61	5	2	15	16

Notes Detail may not sum to totals due to rounding
↑ Significantly higher than the average of countries
↓ Significantly lower than the average of countries

The last two grade four example items were problems set in real-life contexts and involved some algebraic thinking, although they did not necessarily require students to write or solve an equation. The first of these two example items (item 27: Fig. 4.41 and Table 4.36) required students to understand the context, find the cost of a child's ticket, and show their work. The item was worth two score points,

A man took his 3 children to a fair. Tickets cost twice as much for adults as for children. The father paid a total of 50 zeds for the 4 tickets.

How many zeds did each child's ticket cost? Show your work.

Answer: ______________

Item information
Item ID:
M031247

Year(s) administered:
2007, 2003

Performance objective:
Apply algebraic thinking to solve simple real-life problems involving unknowns

Scoring guide

Correct response	
20	10 or 10 zeds with work shown
Partial response	
10	10 or 10 zeds with no work shown
11	Correct method but computation error
Incorrect response	
70	50/4 or 12.5
79	Other incorrect (including crossed out, erased, stray marks, illegible, or off task)
Non response	
99	Blank

Fig. 4.41 TIMSS grade four mathematics item 27, 2007. *Source* TIMSS 2007 Assessment. Copyright © 2009 International Association for the Evaluation of Educational Achievement (IEA). TIMSS & PIRLS International Study Center, Lynch School of Education, Boston College

Table 4.36 Student performance data for mathematics item 27 (M031247), 2007

Country	Percentage of students (%)						
	Correct (weighted)	20	10	11	70	79	99
Italy	17 ⬇	9	16	0	5	47	22
Norway	32 ⬆	17	30	0	1	40	12
Russian Federation	26	12	28	0	2	45	14
Slovenia	20 ⬇	15	9	1	0	66	9
United States	26	21	10	1	11	54	3
Average of countries	24	15	19	0	4	50	12

Notes Correct (weighted) reflects the weighted percentage of students receiving full credit (code 20) or partial credit (code 10 or 11). Detail may not sum to totals due to rounding. Although the table displays rounded data, the calculations of weighted percent correct are based on unrounded data

⬆ Significantly higher than the average of countries

⬇ Significantly lower than the average of countries

and partial credit was given if the answer was correct but no work was shown (code 10 in the scoring guide) or if the response was formulated using the correct method but the answer was not correct due to the student making a computational error (code 11). This was one of the more difficult items for grade four students. On average across the five countries, the weighted percent correct (which takes both full and partial credit into consideration) was only 24, and 66% of students did not receive any credit for this item (codes 70, 79, and 99). Code 70 was given for the responses where students ignored or did not understand the relationship provided for the adult versus child ticket, and simply divided the cost by the number of individuals.

In the last item in the set (item 28: Fig. 4.42 and Table 4.37), two relationships are shown in picture format. Students needed to understand the provided relationships between the cost of ice cream cones and lollipops to solve the problem. In part A, they need to find the cost of one ice cream cone and one lollipop together, and in part B the cost of one lollipop. Similar to the previous item, this was also a difficult item, with 25% correct (weighted) on average across the five countries. On average, 16% of students got both parts A and B correct (code 20 in the scoring guide), 10% got only Part A correct (code 10) and 8% got only part B correct (code 11), resulting in a weighted percent correct of 25%. Performance on this item ranged from 20% correct in Italy to 35% correct in Norway, with both being significantly different from the average across the five countries. Performance in the Russian Federation, Slovenia, and the United States was quite similar to each other and to the average across five countries (22–24%).

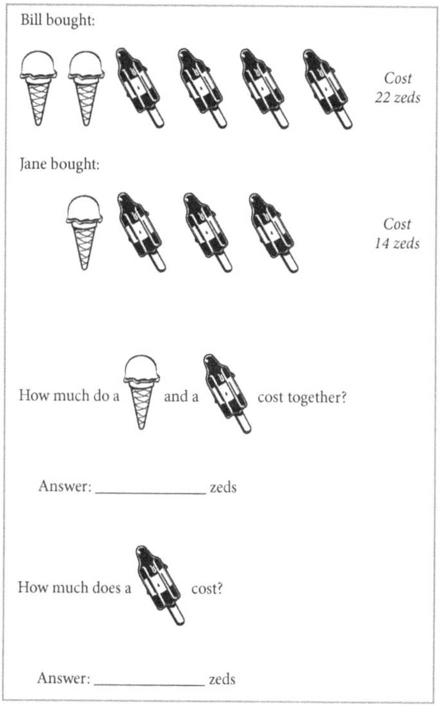

Item information

Item ID:
M051006

Year(s) administered:
2015, 2011

Performance objective:
Apply algebraic thinking to solve simple real-life problems involving unknowns

Scoring guide

Correct response	
20	8 AND 3
Partial response	
10	Only 8 correct
11	Only 3 correct
Incorrect response	
79	Incorrect (including crossed out, erased, stray marks, illegible, or off task)
Non response	
99	Blank

Fig. 4.42 TIMSS grade four mathematics item 28, 2015. *Source* TIMSS 2015 Assessment. Copyright © 2017 International Association for the Evaluation of Educational Achievement (IEA). TIMSS & PIRLS International Study Center, Lynch School of Education, Boston College

Table 4.37 Student performance data for mathematics item 28 (M051006), 2015

Country	Percentage of students (%)					
	Correct (weighted)	20	10	11	79	99
Italy	20 ⬇	13	7	6	60	14
Norway	35 ⬆	14	10	9	54	13
Russian Federation	22	24	15	6	48	7
Slovenia	24	15	7	8	59	11
United States	24	14	8	10	65	2
Average of countries	25	16	10	8	57	9

Notes Correct (weighted) reflects the weighted percentage of students receiving full credit (code 20) or partial credit (code 10 or 11). Detail may not sum to totals due to rounding. Although the table displays rounded data, the calculations of weighted percent correct are based on unrounded data

⬆ Significantly higher than the average of countries

⬇ Significantly lower than the average of countries

4.3.3 Patterns in Errors and Misunderstandings Related to Linear Equations Across Grade Levels and Countries

Student performance data on the individual assessment items described in Sect. 4.3.2 were combined to explore patterns in the percentage of students demonstrating specific errors and misunderstandings across countries, based on the set of items that measure each type of error or misunderstanding at each grade level (Figs. 4.43, 4.44, and 4.45).[22]

The first performance objective is to interpret the solution to a system of linear equations to answer a question or solve a problem in a real-life context. On average across the five countries, 63% of TIMSS Advanced students (Fig. 4.43) demonstrated misunderstanding M1 ("not able to use slope and intercept to provide an argument in support of the solution to a real-life problem situation") on item 1B (Fig. 4.28). In the Russian Federation, Slovenia, and the United States, the percentage of students with this misunderstanding was similar to the average. In Italy, the percentage of students demonstrating this misunderstanding was higher (75%) and, in Norway, it was lower (57%) than the average.

Under performance objective 2 ("solve systems of linear equations in two variables"), there are two types of errors, depending on whether students are applying the procedure to a contextualized real-life problem (M2A) or to a non-contextualized problem (M2B). In a contextual situation, students need an additional piece of understanding to evaluate the situation and write the correct equation. This is not needed in the case of a non-contextual situation. There are two example items

[22]The data shown in Fig. 4.43, 4.44 and 4.45 reflect the most recent assessment year, which differs across the set of items at each grade level (from 1995 to 2015). Table 4.22 shows the most recent assessment year for each item.

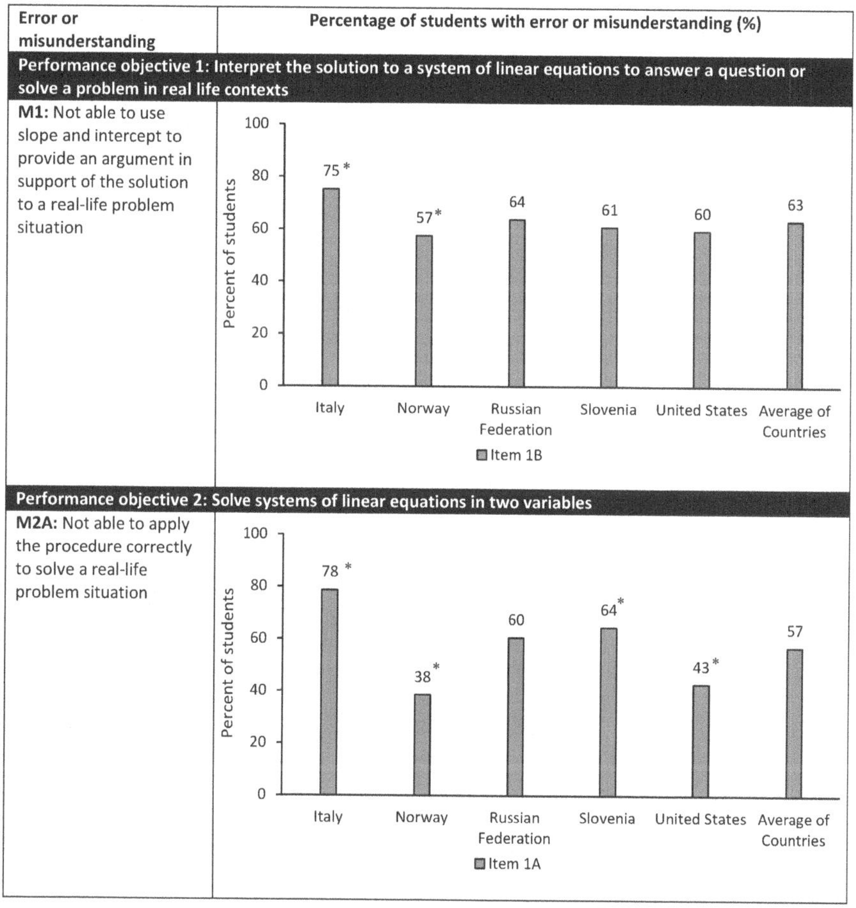

Fig. 4.43 Percentage of TIMSS Advanced students with errors and misunderstandings about linear equations, by country, 2015. *Notes* The percentages are for the most recent cycle each item was administered. Data for item 1A and 1B are from 2015. *Significantly different from average of countries

involving contextualized problems, one from TIMSS Advanced (Fig. 4.28, item 1A) and one from grade eight (Fig. 4.29, item 2). Many students at both grade levels were not able to apply the procedure correctly (error M2A). On average across the five countries, 57% of TIMSS Advanced students and 72% of grade eight students made this error (Figs. 4.43 and 4.44). In Norway, the Russian Federation, and the United States, the percentage of students making error in the procedure was lower in TIMSS Advanced than in grade eight (as might be expected), but in Italy and Slovenia, the percentage of students was comparable at both grade levels. There was more variation across countries in the percentage of students making this error among TIMSS Advanced students (a range of 40 percentage points) than among

Error or misunderstanding	Percentage of students with error or misunderstanding (%)
Performance objective 2: Solve systems of linear equations in two variables	
M2A: Not able to apply the procedure correctly to solve a real-life problem situation	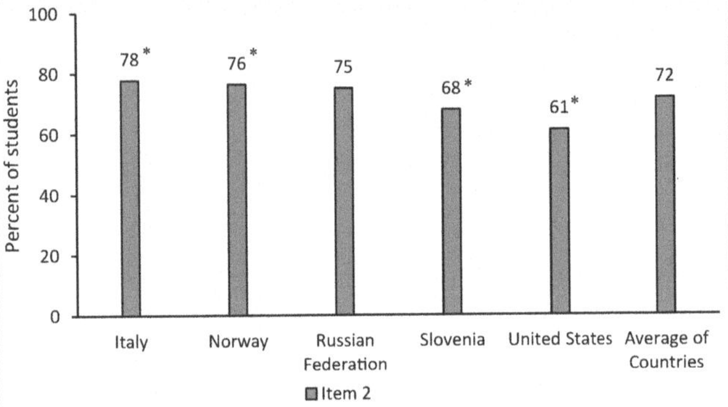
M2B: Not able to apply the procedure correctly to solve non-contextualized problems	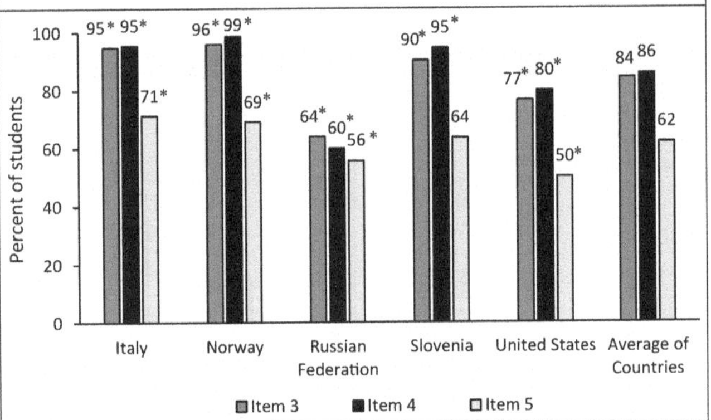
Performance objective 3: Interpret the meanings of slope and y-intercept in linear equations or graphs	
M3A: Not able to relate slope with steepness of a line	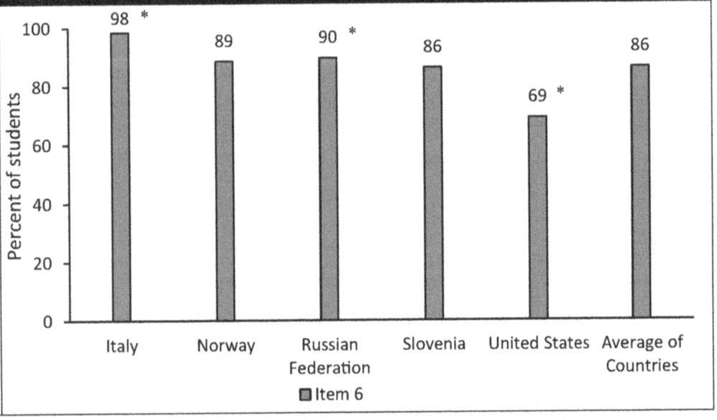

Fig. 4.44 Percentage of grade eight students with errors and misunderstandings about linear equations, by country: 1999, 2003, 2007, and 2015. *Notes* The percentages are for the most recent cycle each item was administered. Data for items 3, 4, 6, 8–14, and 16 are from 2015; data for items 2 and 17 are from 2007; data for items 5, 15 and 20 are from 2003; and data for items 7, 18, and 19 are from 1999. *Significantly different from average of countries. – Data not available (see Appendix for country-specific notes)

Error or misunderstanding	Percentage of students with error or misunderstanding (%)
Performance objective 3: Interpret the meanings of slope and y-intercept in linear equations or graphs	
M3B: Demonstrates confusion between slope and intercept of an equation	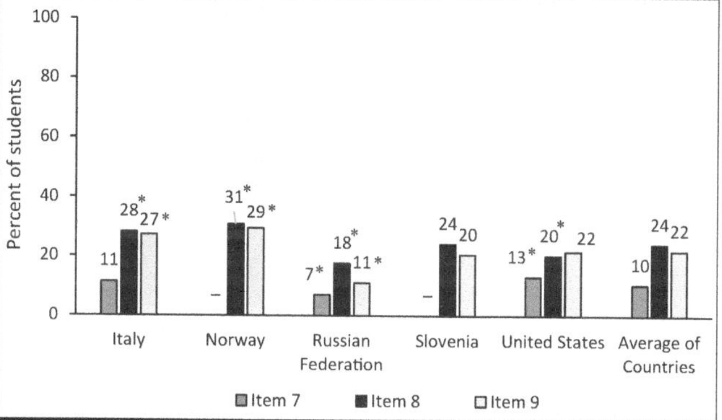
Performance objective 4: Relate algebraic equations to their graphical representations (and vice-versa)	
M4A: Not able to correctly identify the graph of an equation	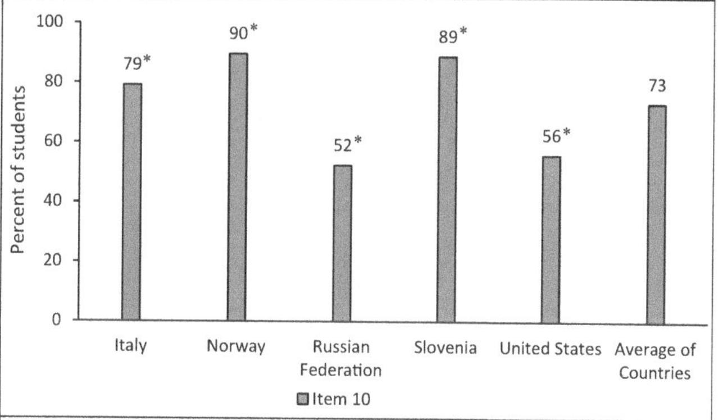
M4B: Not able to translate graphical representations into a mathematical equation or verbal description of a linear relationship	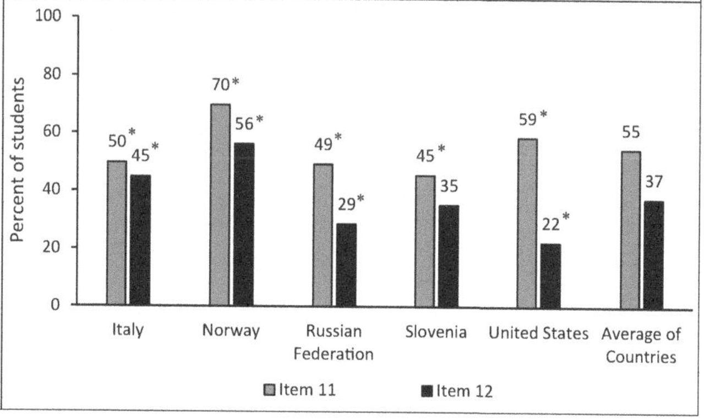

Fig. 4.44 (continued)

Error or misunderstanding	Percentage of students with error or misunderstanding (%)
Performance objective 5: Write equations to represent situations	
M5: Not able to translate verbal descriptions into a correct mathematical equation	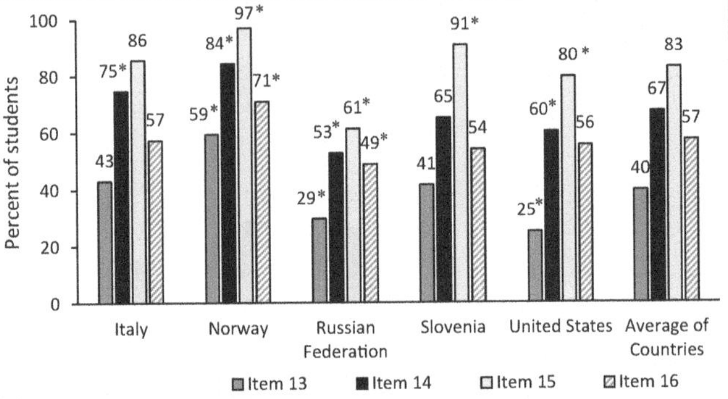
Performance objective 6: Given pairs of numbers in tables or ordered pairs, generate an algebraic equation of the relationship between the two variables	
M6: Not able to translate relationship shown in table form into a mathematical equation	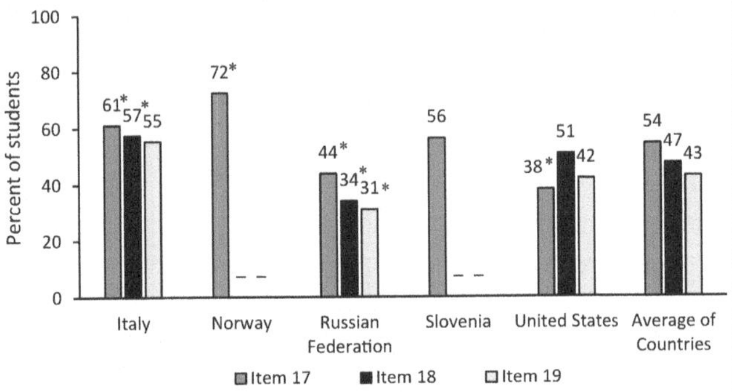
Performance objective 7: Given pairs of numbers in tables or ordered pairs, generate a verbal description of the relationship	
M7A: Not able to generate a correct verbal description given a specific relationship in the form of ordered pairs	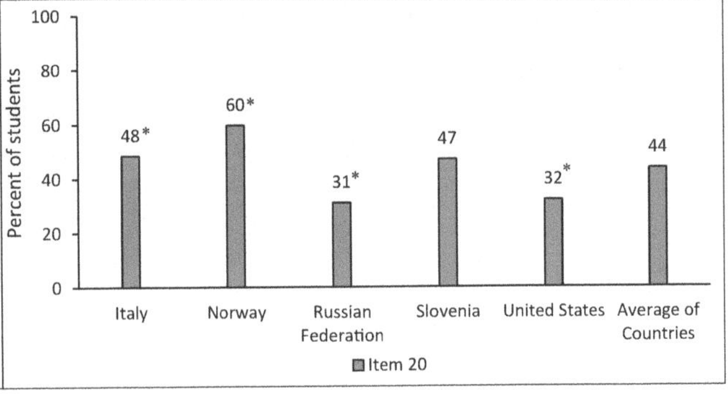

Fig. 4.44 (continued)

Error or misunderstanding	Percentage of students with error or misunderstanding (%)

Performance objective 7: Given pairs of numbers in tables or ordered pairs, generate a verbal description of the relationship

M7B: Not able to generate a correct verbal description given a specific relationship shown in table form

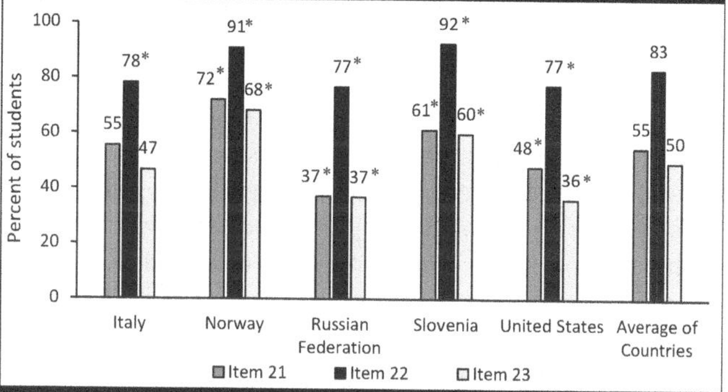

Performance objective 8: Given a verbal description of a relationship between a set of numbers, generate pairs of whole numbers that follow that relationship (rule)

M8: Not able to identify a correct set of numbers that follow a given relationship/rule

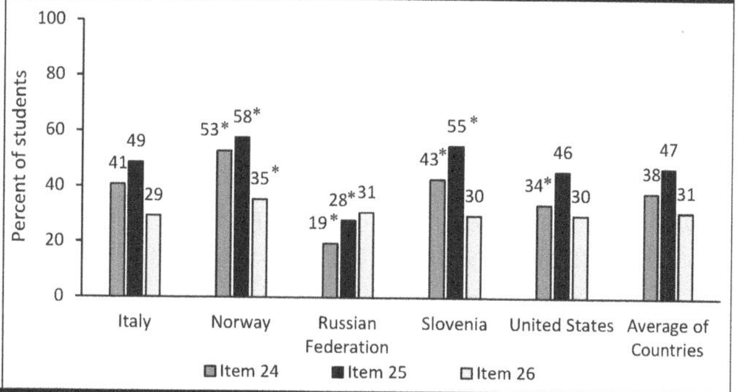

Performance objective 9: Apply algebraic thinking to solve simple real-life problems involving unknowns

M9: Not able to apply algebraic thinking to solve simple real-life problems involving unknowns

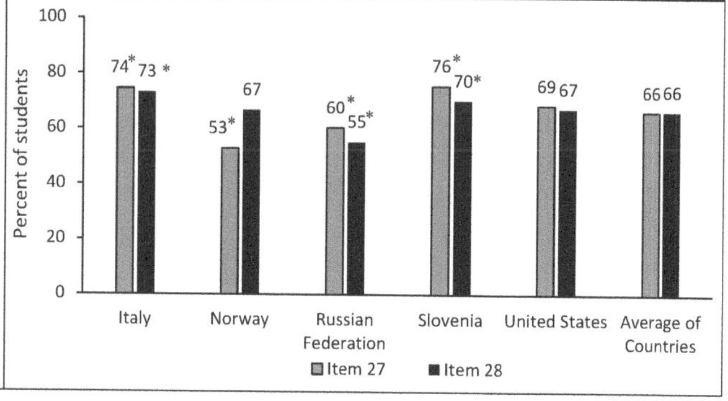

Fig. 4.45 Percentage of grade four students with errors and misunderstandings about linear equations, by country, 2007, 2011, and 2015. *Notes* The percentages are for the most recent cycle each item was administered. Data for items 21, 24, 25, and 28 are from 2015; data for item 23 are from 2011; and data for items 22, 26, and 27 are from 2007. *Significantly different from average of countries

grade eight students (a range of 17 percentage points). In TIMSS Advanced, the percentage of upper-secondary students not able to apply the procedure ranged from 38% of students in Norway and 43% of students in the United States, to 60% in the Russian Federation and 64% in Slovenia, to 78% in Italy.

The related error (M2B) occurred when students were not able to apply the procedure correctly to solve a non-contextualized problem. There were three items at grade eight, item 3 (Fig. 4.30), item 4 (not shown), and item 5 (not shown), that assessed whether students could correctly apply the procedure to solve a system of linear equations (Fig. 4.44). The average percentage of students making errors in applying the procedure was very high across countries on all three items (84%, 86%, and 62% on items 3, 4, and 5, respectively). As expected, this was more pronounced in the CR items (items 3 and 4) than in the MC item (item 5). The pattern across items was very similar in all counties except the Russian Federation. In the case of the Russian Federation, the difference between CR and MC items was less pronounced (64%, 60%, and 56% on items 3, 4, and 5, respectively).

Another important and prevalent misunderstanding among grade eight students is not being able to relate the steepness of a line with the slope of the line (M3A). The percentage of students demonstrating this misunderstanding on the CR item 6 (not shown) was very high in all countries, with four of the five countries having at least 86% of students in this category. In comparison, 69% of students in the United States demonstrated this misunderstanding (Fig. 4.44).

A related misunderstanding that students demonstrate is confusion between the slope and intercept of an equation (M3B). At grade eight, there are three MC items included in this set that are related to this misunderstanding: item 7 (not shown), item 8 (not shown), and item 9 (Fig. 4.31). For all three items, this specific misunderstanding was tracked by having a distractor in which the two values for intercept and slope were swapped in the equations. This misunderstanding was found to be lower among grade eight students than were some of the other misunderstandings and/or errors (ranging from 10 to 24%, on average). This misunderstanding was less common among students in the Russian Federation (demonstrated by 7–18% of students, depending on the item), but more common in Italy (11–28%), Norway (29–31%), Slovenia (20–24%), and the United States (13–22%).

Performance objective 4 is related to students being able to translate easily between an algebraic equation and the graph of a line. The first misunderstanding related to this performance objective is that students are not able to correctly identify the graph of an equation (M4A). More than 75% of students in Italy, Norway, and Slovenia demonstrated this misunderstanding and were not able to identify the correct graph for a given equation on item 10 (not shown).

A related error is not being able to write or identify the correct algebraic equation or verbal description from the graph of a line (M4B). Students found translating the graph of a line to its algebraic form (item 11: Fig. 4.32) more difficult than translating the graph of a line to a verbal description of the relationship on item 12 (not shown). On average across the five countries, 55% of students were not able to select the correct equation for the given graph of a line (item 11), and 37% of students were not able to select the correct description of the relationship/rule for the given graph of a line (item 12). The percentage of students demonstrating the error (M4B) on item 11 varied from 70% in Norway to 45% in Slovenia. In contrast, students demonstrating the error (M4B) on item 12 varied from 56%, again in Norway, to 22% in the United States. This means students understood the relationship between the two variables but found it difficult to verbalize the relationship in algebraic form. The difference between the error demonstrated on items 11 and 12 was most pronounced in the Russian Federation (20 percentage points) and the United States (37 percentage points) and least pronounced in Italy (5 percentage points).

The next type of misunderstanding (M5) was not being able to translate verbal descriptions into a correct mathematical equation. There were four grade eight items measuring this error: two MC (item 13 in Fig. 4.33 and item 16, not shown) and two CR (items 14 and 15, not shown). As was found with other errors and misunderstandings, the percentage of students demonstrating the misunderstanding was higher on the CR items than on the MC items in all five countries. On all four items, the percentage of students demonstrating the misunderstanding was highest in Norway and lowest in the Russian Federation (except for item 13). On item 13, the percentage of students demonstrating the misunderstanding was lowest in United States.

The related misunderstanding (M6) was not being able to translate the relationship given in a table format into a linear equation. All three grade eight items assessing this misunderstanding were MC in format: item 17 (Fig. 4.34), item 18 (now shown), and item 19 (Fig. 4.35).[23] Across all three items, the misunderstanding was more common among students in Italy than among students in the Russian Federation and the United States. Another related misunderstanding (M7A) is not being able to generate a verbal description given a specific relationship in the form of ordered pairs, which was measured on one grade eight item. The percentage of grade eight students demonstrating this error on item 20 was highest in Norway (60%) and lowest in the Russian Federation and the United States (31% and 32%, respectively).

The next three kinds of errors and misunderstandings were demonstrated at grade four (Fig. 4.45). The first one (M7B) is not being able to generate a verbal description

[23]Norway and Slovenia did not participate in the 1999 and 1995 assessments.

given a relationship in table format. There are three items included in this set (items 21, 22, and 23). Item 21 (Fig. 4.37) and item 23 (not shown) are MC format, and item 22 (Fig. 4.38) is CR format. The misunderstanding appears to be more prevalent on the CR item than on the MC items for all countries. The misunderstanding was more common in Norway (68–91% of students across items) and Slovenia (60–92%) than in Italy (47–78%), the Russian Federation (37–77%), and the United States (36–77%).

Another related error (M8) was not being able to identify a correct set of numbers based on the verbal description of the relationship. There were three grade four CR items that measured this misunderstanding: item 24 (Fig. 4.39), item 25 (not shown), and item 26 (Fig. 4.40). The percentage of students demonstrating this type of error on item 24 covered a range of 34 percentage points across the five countries, from 19% in the Russian Federation to 53% in Norway. Similarly, the percentage of students demonstrating this type of error for item 25 covered a range of 30 percentage points, from 28% in the Russian Federation to 58% in Norway. For item 26, the percentage of students demonstrating this error was more consistent across countries, ranging from 29% in Italy to 35% in Norway.

The last type of error or misunderstanding (M9) was that students were not able to apply algebraic thinking to solve simple real-life problems, which is a precursor skill for linear equations. Both example items 27 (Fig. 4.41) and 28 (Fig. 4.42) were CR in format. In responding to these items, students were not expected to formally write equations but to apply algebraic thinking to solve them. Two-thirds of students on average across the five countries (and at least half in each country) were not able to solve these problems correctly.

4.3.4 Gender Differences in Errors and Misunderstandings Related to Linear Equations

On average across the five countries, there were not many significant gender differences found on the set of items related to linear equations. Patterns in the percent correct by gender (Table 4.38)[24] and the percentage of students with specific errors and misunderstanding by gender (Tables 4.39, 4.40, and 4.41, and Figs. 4.46, 4.47, and 4.48)[25] differed across countries and grade levels. Of the few gender differences observed for linear equation items, more favored males than females.

[24]Table 4.38 displays the percent correct for female and male students for each mathematics item. The corresponding percentages correct for all students are shown in Fig. 4.25, 4.26, and 4.27.

[25]Tables 4.39, 4.40 and 4.41 display the percentage of female and male students with each error or misunderstanding in TIMSS Advanced, grade eight, and grade four, respectively. The accompanying figures (Figs. 4.46, 4.47, and 4.48) provide graphical displays of the differences in the percentage of female and male students at the corresponding grade level. The corresponding percentages of students overall with the errors or misunderstandings are shown in Figs. 4.43, 4.44, and 4.45.

Table 4.38 Performance of female and male students on TIMSS and TIMSS Advanced mathematics items, by country and grade level, 1999–2015

Item	Year	Percentage of students correct (%)											
		Italy		Norway		Russian Federation		Slovenia		United States		Average of countries	
		Female	Male	Female	Male	Female	Male	Female	Male	Female	Male	Female	Male
TIMSS Advanced													
Item 1A	2015	16	25	60	62	34	45	28	47	61	54	39	47
Item 1B	2015	17	29	43	42	29	44	32	49	41	40	32	41
TIMSS grade 8													
Item 2	2007	20	19	19	18	22	17	29	32	38	37	25	25
Item 3	2015	3	7	4	4	33	39	7	12	21	26	14	18
Item 4	2015	3	6	0	1	34	38	3	6	19	16	12	13
Item 5	2003	31	27	31	31	44	45	39	34	51	48	39	37
Item 6	2015	2	1	13	10	7	13	14	13	30	32	13	14
Item 7	1999	50	40	–	–	59	56	–	–	58	52	56	49
Item 8	2015	40	40	42	27	58	63	47	43	65	63	50	47
Item 9	2015	39	39	24	19	59	63	39	30	58	53	44	41
Item 10	2015	21	21	9	11	53	43	8	14	43	45	27	27
Item 11	2015	51	50	33	28	53	49	54	55	42	41	47	44
Item 12	2015	51	59	39	49	66	77	67	62	79	76	61	65
Item 13	2015	58	57	42	39	67	74	60	57	78	72	61	60
Item 14	2015	23	27	16	16	50	45	35	35	40	39	33	32
Item 15	2003	14	15	3	3	44	33	10	8	20	20	18	16
Item 16	2015	45	41	30	28	53	50	49	43	46	43	45	41
Item 17	2007	33	45	27	28	56	57	45	43	63	61	45	46
Item 18	1999	45	40	–	–	67	65	–	–	48	51	53	52
Item 19	1999	45	45	–	–	74	64	–	–	58	58	59	56
Item 20	2003	50	53	42	39	67	71	58	48	67	70	57	56
TIMSS grade 4													
Item 21	2015	41	48	25	30	66	60	37	40	51	53	44	46
Item 22	2007	21	23	9	9	26	21	6	9	22	23	17	17
Item 23	2011	54	53	26	37	65	62	38	42	63	65	49	52
Item 24	2015	59	59	48	47	82	79	61	54	70	63	64	60
Item 25	2015	50	53	39	45	74	70	42	48	54	55	52	54
Item 26	2007	62	66	47	61	67	60	59	62	63	62	60	62
Item 27	2007	15	20	33	32	27	25	17	23	27	26	24	25
Item 28	2015	16	24	20	27	33	36	21	24	23	24	23	27

Key

- Higher percentage (%) of females with item correct
- Higher percentage (%) of males with item correct
- No significant difference between females and males

Notes Percent correct is the percentage of students receiving credit on each item. For MC and short CR items (each worth one score point), this reflects the percentage of students who provided a correct answer. For extended CR items, this reflects the weighted percentage of students receiving full credit (2 points) or partial credit (1 point); Item 1 (parts A and B) was scored using an overall scoring guide (shown in Fig. 4.28). The percent correct shown for 1A reflects all students who answered part A correctly (codes 20 and 10 combined). The percent correct shown for 1B reflects all students who answered part B correctly (codes 20 and 11 combined)

– Data not available (see Appendix for country-specific notes)

Table 4.39 Percentage of female and male TIMSS Advanced students with errors and misunderstandings about linear equations, by country, 2015

Errors and misunderstandings	Item	Year	Percentage of students with error or misunderstanding (%)											
			Italy		Norway		Russian Federation		Slovenia		United States		Average of countries	
			Female	Male	Female	Male	Female	Male	Female	Male	Female	Male	Female	Male
Performance objective 1: Interpret the solution to a system of linear equations to answer a question or solve a problem in real life contexts														
M1: Not able to use slope and intercept to provide an argument in support of the solution to a real-life problem situation	Item 1B	2015	83	71	57	58	71	56	68	51	59	60	68	59
Performance objective 2: Solve systems of linear equations in two variables														
M2A: Not able to apply the procedure correctly to solve a real-life problem situation	Item 1A	2015	84	75	40	38	66	55	72	53	39	46	61	53

Key

- (dark shading) Higher percentage (%) of females with error or misunderstanding
- (light shading) Higher percentage (%) of males with error or misunderstanding
- (no shading) No significant difference between females and males

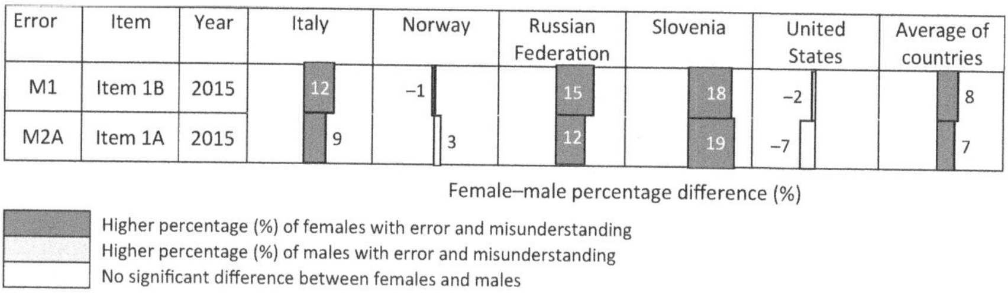

Fig. 4.46 Gender differences in errors and misunderstandings about linear equations among TIMSS Advanced students, 2015. *Notes* Mathematics errors and misunderstandings: M1 = not able to use slope and intercept to provide an argument in support of the solution to a real-life problem situation, M2A = not able to apply the procedure correctly to solve a real-life problem situation

Gender differences in percent correct were greatest on the TIMSS Advanced items, with an average female-male difference of at least 8% in favor of boys on both items (Table 4.38). However, this varied across countries, with significant differences found in Italy, the Russian Federation, and Slovenia, but not in Norway or the United States. In comparison, on average across the 19 items at grade eight, significant female-male differences in item percent correct ranged from 4–7%, with two items having measurably different performance that favored boys and two items that favored girls. In all countries, there were two or three grade eight items with significant gender differences. On average at grade four, only one of the eight items had a significant item performance difference that favored males (by 4%). However, in Norway, there were three items where the percent correct favored males.

The specific set of items with significant gender differences varied across countries. None of the items had significant gender differences in all five countries, but the two TIMSS Advanced items had significant gender differences in three countries. In contrast, 12 of the 19 items at grade eight and five of the eight items at grade four had significant gender differences in one or two countries.

Looking at the percent correct in each country, Italy had significant gender differences that favored males on both items in TIMSS Advanced and one item each at grade eight (item 17) and grade four (item 28); grade eight females in Italy performed better on one item (item 7). Norway had one item at grade eight (item 12) and three items at grade four (items 23, 26, and 28) where males performed better than females, and one item at grade eight (item 8) where females performed better than males. The Russian Federation had significant gender differences in favor of males on both items in TIMSS Advanced. In grade eight, there were two items (items 6 and 12) where males performed better than females and two items (items 15 and 19) where females performed better than males. There were no significant gender differences on any grade four items in the Russian Federation.

Slovenia had both items in TIMSS Advanced, two items at grade eight (items 3 and 10), and one item at grade four (item 27) where males performed better than females, and one item at grade eight (item 20) where females performed better than males. In comparison, the United States was the only country with no items where male performance was higher; there were two items at grade eight (items 7 and 13) and one item at grade four (item 24) where females did better than males.

Gender differences in the percentage of students demonstrating errors or misunderstandings were greatest for the TIMSS Advanced items (Table 4.39 and Fig. 4.46), with an average female-male difference of 8–9% on items 1A and 1B, ranging from 1 to 19% across countries. Significant differences were found on both items in Italy, the Russian Federation, and Slovenia. In all of these cases, there were higher percentages of females than males with these errors or misunderstandings. There were no statistically significant gender differences on either item in Norway and the United States.

At grade eight (Table 4.40 and Fig. 4.47), 19 items were included in the set. For most of these items, the gender differences in the percentage of students with errors and misunderstandings were found to be not significant. For each country, the items exhibiting a gender difference varied from one item in Norway to five items in the Russian Federation. With the exception of items 7 and 12, gender differences on all other items were found in only one country. For item 7, there were greater percentages of male than female students with the misunderstanding in Italy (a difference of 5%) and the Russian Federation (a difference of 4%). For item 12 (error M4B), there was a greater percentage of females than males with the misconception in Norway and the Russian Federation (percentage differences of 10% and 11%, respectively).

The United States was the only country with significant gender differences on items 8 and 13 (misunderstanding/error M3B and M5), with both items having a higher percentage of males with the misunderstanding/error (5–6%). In contrast, the Russian Federation exhibited gender differences on five of the 19 items in the set. Two items (6 and 12) showed a higher percentage of females in the Russian Federation demonstrating the misunderstanding/error (M3A and M4B), and three items (7, 15, and 19) showed a higher percentage of males demonstrating the misconceptions assessed by these items (M3B, M5, and M6).

Italy exhibited gender differences for two of the 19 items. On item 7, 5% more males than females had the misunderstanding (M3B), while on item 17, 12% more females had the misunderstanding (M6). In Norway, only item 12 had a significant difference, with 10% more females than males demonstrating the error (M4B).

In grade four there were few occurrences of significant gender differences related to errors and misunderstandings across items and countries (Table 4.41 and Fig. 4.48). In Italy and the Russian Federation, there were no significant gender differences on any of the eight items included in the set at grade four. Both Slovenia and the United States had one item each with significant gender differences. In the case of Slovenia, there was a higher percentage of females than males on item 27 (9%) with misunderstanding M9. In contrast, in the United States, a higher percentage of males than females (7%) demonstrated misunderstanding M8 on item 24.

Table 4.40 Percentage of female and male grade eight students with errors and misunderstandings about linear equations, by country: 1999, 2003, 2007, and 2015

Errors and misunderstandings	Item	Year	Percentage of students with error or misunderstanding (%)											
			Italy		Norway		Russian Federation		Slovenia		United States		Average of countries	
			Female	Male	Female	Male	Female	Male	Female	Male	Female	Male	Female	Male
Performance objective 2: Interpret the solution to a system of linear equations to answer a question or solve a problem in real life contexts														
M2A: Not able to apply the procedure correctly to solve a real-life problem situation	Item 2	2007	77	78	76	77	73	77	71	65	62	61	72	71
M2B: Not able to apply the procedure correctly to solve non-contextualized problems	Item 3	2015	97	93	96	96	67	61	93	88	79	74	86	82
	Item 4	2015	97	94	99	98	61	59	96	93	78	82	86	85
	Item 5	2003	69	73	69	69	56	55	61	66	49	52	61	63
Performance objective 3: Interpret the meanings of slope and y-intercept in linear equations or graphs.														
M3A: Not able to relate slope with steepness of a line	Item 6	2015	98	99	87	90	93	87	86	87	70	68	87	86
M3B: Demonstrates confusion between slope and intercept of an equation	Item 7	1999	9	14	–	–	5	9	–	–	12	14	9	12
	Item 8	2015	27	29	30	31	18	17	24	25	18	23	24	25
	Item 9	2015	26	28	26	33	12	10	21	20	22	21	21	22
Performance objective 4: Relate algebraic equations to their graphical representations (and vice-versa).														
M4A: Not able to correctly identify the graph of an equation	Item 10	2015	79	79	91	89	47	57	92	86	57	55	73	73
M4B: Not able to translate graphical representations into a mathematical equation or verbal description of a linear relationship	Item 11	2015	49	50	67	72	47	51	46	45	58	59	53	56
	Item 12	2015	49	41	61	51	34	23	33	38	21	24	39	35

(continued)

(continued)

Errors and misunderstandings	Item	Year	Percentage of students with error or misunderstanding (%)											
			Italy		Norway		Russian Federation		Slovenia		United States		Average of countries	
			Female	Male	Female	Male	Female	Male	Female	Male	Female	Male	Female	Male
Performance objective 5: Write equations to represent situations														
M5: Not able to translate verbal descriptions into a correct mathematical equation	Item 13	2015	42	43	58	61	33	26	40	43	22	28	39	40
	Item 14	2015	77	73	84	84	50	55	65	65	60	61	67	68
	Item 15	2003	86	85	97	97	56	67	90	92	80	80	82	84
	Item 16	2015	55	59	70	72	47	50	51	57	54	57	55	59
Performance objective 6: Given pairs of numbers in tables or ordered pairs, generate an algebraic equation of the relationship between the two variables														
M6: Not able to translate relationship shown in table form into a mathematical equation	Item 17	2007	67	55	73	72	44	43	55	57	37	39	55	54
	Item 18	1999	55	60	–	–	33	35	–	–	52	49	47	48
	Item 19	1999	55	55	–	–	26	36	–	–	42	42	41	44
Performance objective 7: Given pairs of numbers in tables or ordered pairs, generate a verbal description of the relationship														
M7A: Not able to generate a correct verbal description given a specific relationship in the form of ordered pairs	Item 20	2003	50	47	58	61	33	29	42	52	33	30	43	44

Key

- (dark shading) Higher percentage (%) of females with error or misunderstanding
- (light shading) Higher percentage (%) of males with error and misunderstanding
- (no shading) No significant difference between females and males

Notes

– Data not available (see Appendix for country-specific notes)

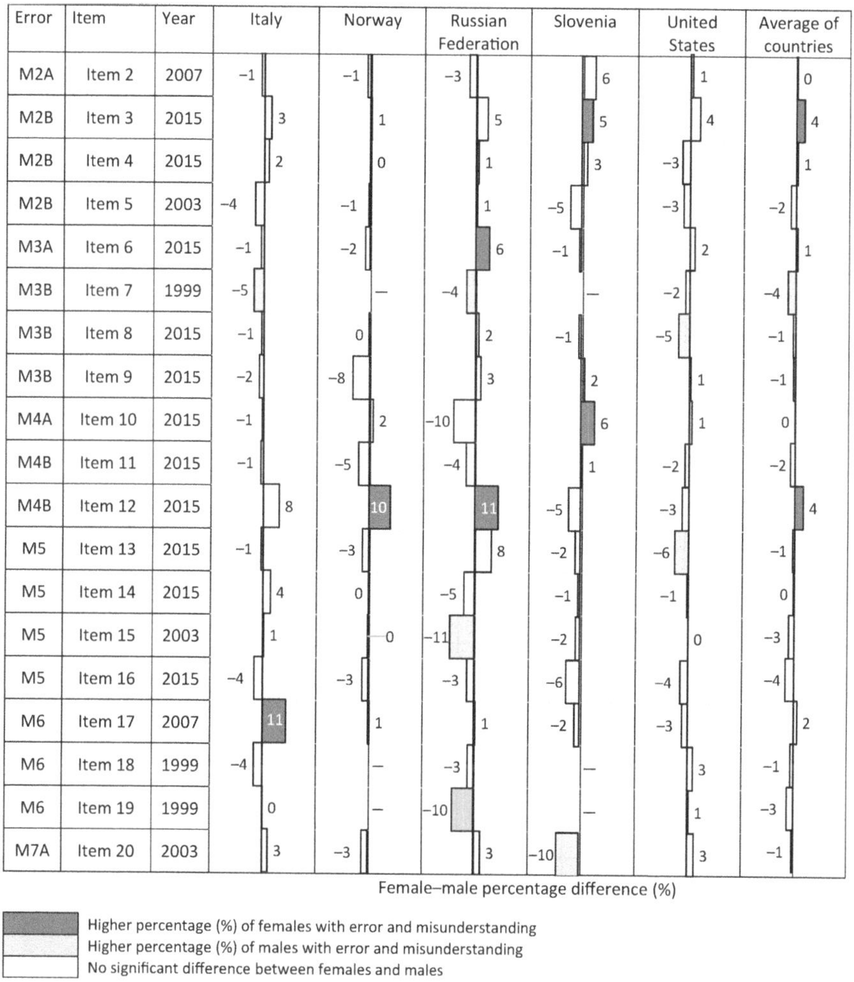

Fig. 4.47 Gender differences in errors and misunderstandings about linear equations among grade eight students, 1999, 2003, 2007, and 2015. *Notes* Mathematics errors and misunderstandings: M2A = not able to apply the procedure correctly to solve a real-life problem situation, M2B = not able to apply the procedure correctly to solve non-contextualized problems, M3A = not able to relate slope with steepness of a line, M3B = demonstrates confusion between slope and intercept of an equation, M4A = not able to correctly identify the graph of an equation, M4B = not able to translate graphical representations into a mathematical equation or verbal description of a linear relationship, M5 = not able to translate verbal descriptions into a correct mathematical equation, M6 = not able to translate relationship shown in table form into a mathematical equation, M7A = not able to generate a correct verbal description given a specific relationship in the form of ordered pairs. – Data not available (see Appendix for country-specific notes)

Table 4.41 Percentage of female and male grade four students with errors and misunderstandings about linear equations, by country, 2007, 2011, and 2015

Errors and misunderstandings	Item	Year	Percentage of students with error or misunderstanding (%)											
			Italy		Norway		Russian Federation		Slovenia		United States		Average of countries	
			Female	Male	Female	Male	Female	Male	Female	Male	Female	Male	Female	Male
Performance objective 7: Given pairs of numbers in tables or ordered pairs, generate a verbal description of the relationship														
M7B: Not able to generate a correct verbal description given a specific relationship shown in table form	Item 21	2015	59	52	75	70	34	40	63	60	49	47	56	54
	Item 22	2007	79	77	91	91	74	79	94	91	78	77	83	83
	Item 23	2011	46	47	74	63	35	38	62	58	37	35	51	48
Performance objective 8: Given a verbal description of a relationship between a set of numbers, generate pairs of whole numbers that follow that relationship (rule)														
M8: Not able to identify a correct set of numbers that follow a given relationship/rule	Item 24	2015	41	41	52	53	18	21	39	46	30	37	36	40
	Item 25	2015	50	47	61	55	26	30	58	52	46	45	48	46
	Item 26	2007	33	26	43	29	28	34	32	28	29	31	33	29
Performance objective 9: Apply algebraic thinking to solve simple real-life problems involving unknowns														
M9: Not able to apply algebraic thinking to solve simple real-life problems involving unknowns	Item 27	2007	78	71	54	52	58	62	80	71	69	68	68	65
	Item 28	2015	77	69	70	63	57	53	73	68	68	67	69	64

Key

- (dark shading) Higher percentage (%) of females with error or misunderstanding
- (light shading) Higher percentage (%) of males with error or misunderstanding
- (no shading) No significant difference between females and males

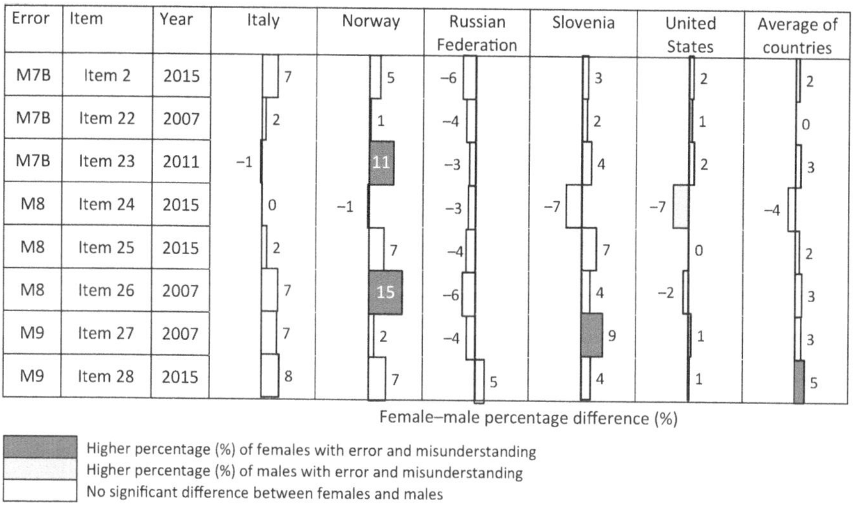

Fig. 4.48 Gender differences in errors and misunderstandings about linear equations among grade four students, 2007, 2011, and 2015. *Notes* Mathematics errors and misunderstandings: M7B = not able to generate a correct verbal description given a specific relationship shown in table form, M8 = not able to identify a correct set of numbers that follow a given relationship/rule, M9 = not able to apply algebraic thinking to solve simple real-life problems involving unknowns

Norway is the only country with a significant gender difference on two items (items 23 and 26) showing a higher percentage of females (11% and 14%, respectively) demonstrating the misunderstandings (M7B and M8).

4.3.5 *Patterns in Errors and Misunderstandings Related to Linear Equations Over Time*

In this section, we present the percentage of students in each country demonstrating a specific type of error or misunderstanding over multiple assessment years for the set of trend items at each grade level (Figs. 4.49 and 4.50). For the linear equations topic, there were 10 trend items at grade eight and seven trend items at grade four, but no trend items available for TIMSS Advanced. At both grades four and eight, there were three items administered in three assessment years before they were released, and all other items were administered for two assessment cycles.

Looking at grade eight (Fig. 4.49), the trend item data covered assessment years 1995, 1999, 2003, 2007, 2011, and 2015. There are some significant differences across assessment years in the percentage of students in each country demonstrating the specific types of errors or misunderstandings. Item 4 (administered in 2011 and 2015) measures error M2B ("not able to apply the procedure correctly to solve non-contextualized problems"). This item shows a decrease of 4% of students

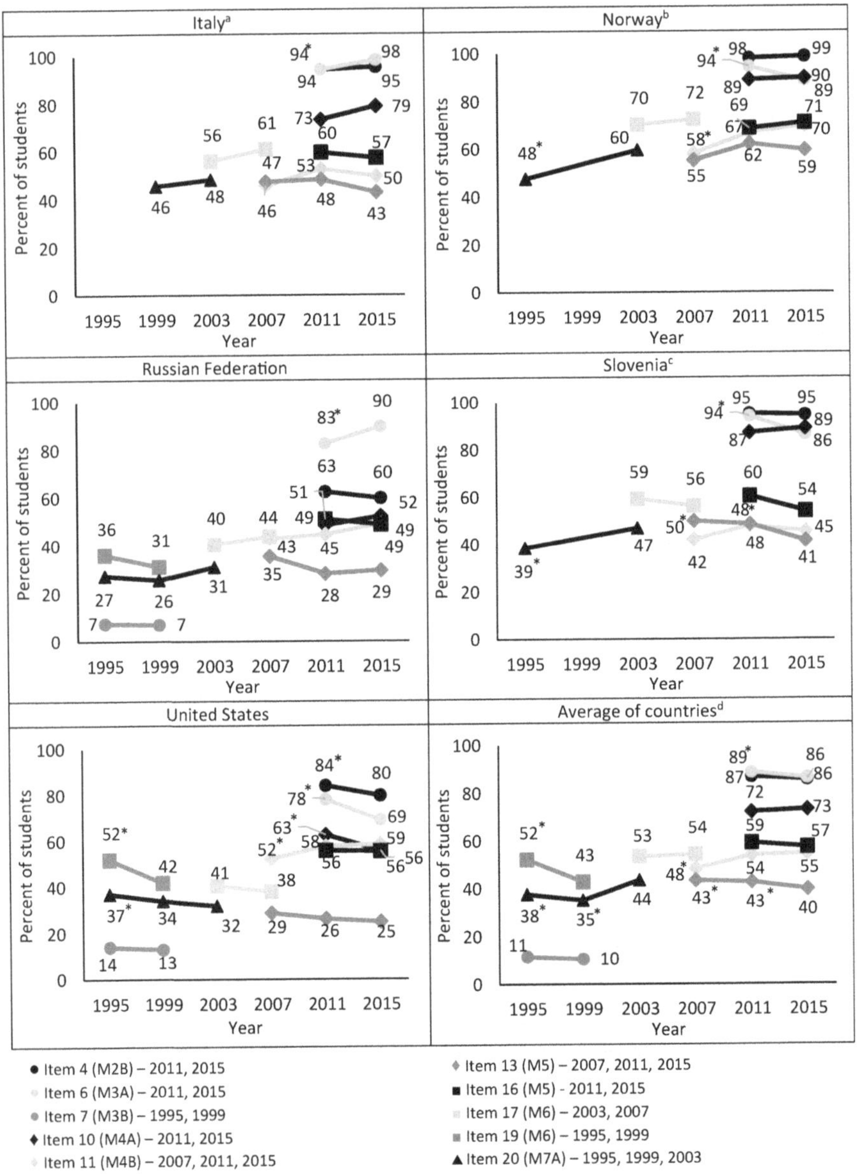
Italy[a]
Norway[b]
Russian Federation
Slovenia[c]
United States
Average of countries[d]
Percent of students
Year
1995 1999 2003 2007 2011 2015
● Item 4 (M2B) – 2011, 2015
Item 6 (M3A) – 2011, 2015
● Item 7 (M3B) – 1995, 1999
♦ Item 10 (M4A) – 2011, 2015
Item 11 (M4B) – 2007, 2011, 2015
♦ Item 13 (M5) – 2007, 2011, 2015
■ Item 16 (M5) - 2011, 2015
Item 17 (M6) – 2003, 2007
■ Item 19 (M6) – 1995, 1999
▲ Item 20 (M7A) – 1995, 1999, 2003

◂**Fig. 4.49** Trends in the percentage of grade eight students with errors and misunderstandings about linear equations, 1995–2015. *Notes* Mathematics errors and misunderstandings: M2A = not able to apply the procedure correctly to solve a real-life problem situation, M2B = not able to apply the procedure correctly to solve non-contextualized problems, M3A = not able to relate slope with steepness of a line, M3B = demonstrates confusion between slope and intercept of an equation, M4A = not able to correctly identify the graph of an equation, M4B = not able to translate graphical representations into a mathematical equation or verbal description of a linear relationship, M5 = not able to translate verbal descriptions into a correct mathematical equation, M6 = not able to translate relationship shown in table form into a mathematical equation, M7A = not able to generate a correct verbal description given a specific relationship in the form of ordered pairs. *Significantly different from most recent assessment cycle. [a]1995 trend data for Italy are not available for items 7, 19, and 20. [b]1999 trend data for Norway are not available for items 7, 19, and 20. [c]1999 trend data for Slovenia are not available for items 7, 19, and 20. [d]The average of countries for each cycle is calculated using all countries that participated in a given year. Because not all countries have data for each cycle of TIMSS, the average for countries for each year may include a different set of countries and is therefore not directly comparable to other years. In some cases, countries have data for only one year, so their data are not shown in the trend graphs for individual countries, but their data are included in the average of countries for that particular year (see Appendix for country-specific notes)

making the error in the United States in 2015 in comparison to the previous assessment cycle in 2011.

Item 6 (administered in 2011 and 2015) measures misunderstanding M3A ("not able to relate slope with steepness of lines"), and the trend data for this item show an increase in the percentage of students demonstrating this misunderstanding over time for students in Italy (4%) and the Russian Federation (7%), but a decrease for the United States (9%), Norway (5%), and Slovenia (8%). Item 7 (administered in 1995 and 1999) measures the related misunderstanding M3B ("demonstrates confusion between slope and intercept of an equation"). Trend data are available for only two countries, the Russian Federation and the United States. The percentage of students demonstrating this misunderstanding did not change between 1995 and 1999 for either country.

Item 10 (administered in 2011 and 2015) measures error M4A ("not able to correctly identify the graph of an equation"). The percentage of students making this error decreased by 7% in the United States. Another related error M4B ("not able to translate graphical representations into a mathematical equation or verbal description of a linear relationship") was measured by item 11 (administered in 2007, 2011, and 2015). The trend data for this item show no change in the percentage of students demonstrating the error from 2011 for all countries. However, for Norway and the United States, the percentage of students demonstrating the misconception increased after the 2007 cycle of TIMSS.

Item 13 (administered in 2007, 2011, and 2015) and item 16 (administered in 2011 and 2015) both measure error M5 ("not able to translate verbal descriptions into a correct mathematical equation"). For item 13 ("formula for K the cost of trip"), the general trend was a decrease in the percentage of students demonstrating this error in 2015 from the previous two assessment years (based on the average percentage across the five countries). However, in Italy and Norway, the percentage appeared to increase between 2007 and 2011 and then decrease between 2011 and 2015 (though the differences were not statistically significant). For item 16 ("set up system of equations"), the difference in the percentage of students did not change between 2011 and 2015.

Item 17 (administered in 2003 and 2007) and item 19 (administered in 1995 and 1999) both measure misunderstanding M6 ("not able to translate relationship shown in table form into a mathematical equation"). Trend data for item 17 show no statistically significant difference in the percentage of students demonstrating this misunderstanding from 2003 to 2007. For item 19, trend data were only available for the Russian Federation and the United States. On this item, the percentage of students demonstrating the misunderstanding decreased between 1995 and 1999 by 10% in the United States and was not statistically different in the Russian Federation.

Item 20 (administered in 1995, 1999, and 2003) measures misunderstanding M7A ("not able to generate a correct verbal description given a specific relationship in the form of ordered pairs"). For this item, complete data for three assessment cycles were only available for the Russian Federation and the United States. In contrast, data for Italy are available for 1999 and 2003 and data for Norway and Slovenia are available for 1995 and 2003. The available data show that the percentage of students with this misunderstanding increased over time in Norway (by 12%) and Slovenia (8%), but decreased in the United States (by 5%).

At grade four, the trend item data covered assessment years 2003, 2007, 2011, and 2015 (Fig. 4.50). Again, at grade four, there were some significant differences over time in the percentage of students demonstrating the misunderstandings. Item 21 (administered in 2007, 2011, and 2015), item 22 (administered in 2003 and 2007), and item 23 (administered in 2003, 2007, and 2011) all measure misunderstanding M7B ("not able to generate a correct verbal description given a specific relationship shown in table form"). In general, across countries the data show a decrease in students having this misunderstanding. Trend data for item 21 show a significant decrease in the percentage of Slovenian students showing the misunderstanding from 2007 to 2015 (by 7%). On item 22, the percentage of students having this misunderstanding decreased from 2003 to 2007 in the Russian Federation (by 15%) and the United States (by 7%). Similarly, in the case of item 23, the percentage of students having this misunderstanding decreased significantly between 2003 and 2011, by 8% in Italy, and by 11% in the Russian Federation and the United States.

Item 24 (administered in 2007, 2011, and 2015) and item 26 (administered in 2003 and 2007) measure misunderstanding M8 ("not able to identify a correct set of numbers that follow a given relationship/rule"). Trend data for item 24 show some interesting patterns over these three points in time. The percentage of students showing this misunderstanding decreased consistently over time in Norway (by 12%), the Russian Federation (16%), and Slovenia (7%). Trend data for item 26 show that the percentage of students with this misconception generally decreased from 2003 to 2007, with significant decreases in Slovenia (by 13%) and the Russian Federation (14%).

Item 27 (administered in 2003 and 2007) and item 28 (administered in 2011 and 2015) measure misunderstanding M9 ("not able to apply algebraic thinking to solve simple real-life problems involving unknowns"). Trend data for item 27 showed that the percentage of students demonstrating the misunderstanding decreased from 2003 to 2007 in the Russian Federation (by 11%) and was not significantly different over time in the other countries. Data for item 28 showed that the percentage of students with this misconception decreased from 2011 to 2015 only for Slovenia (by 6%).

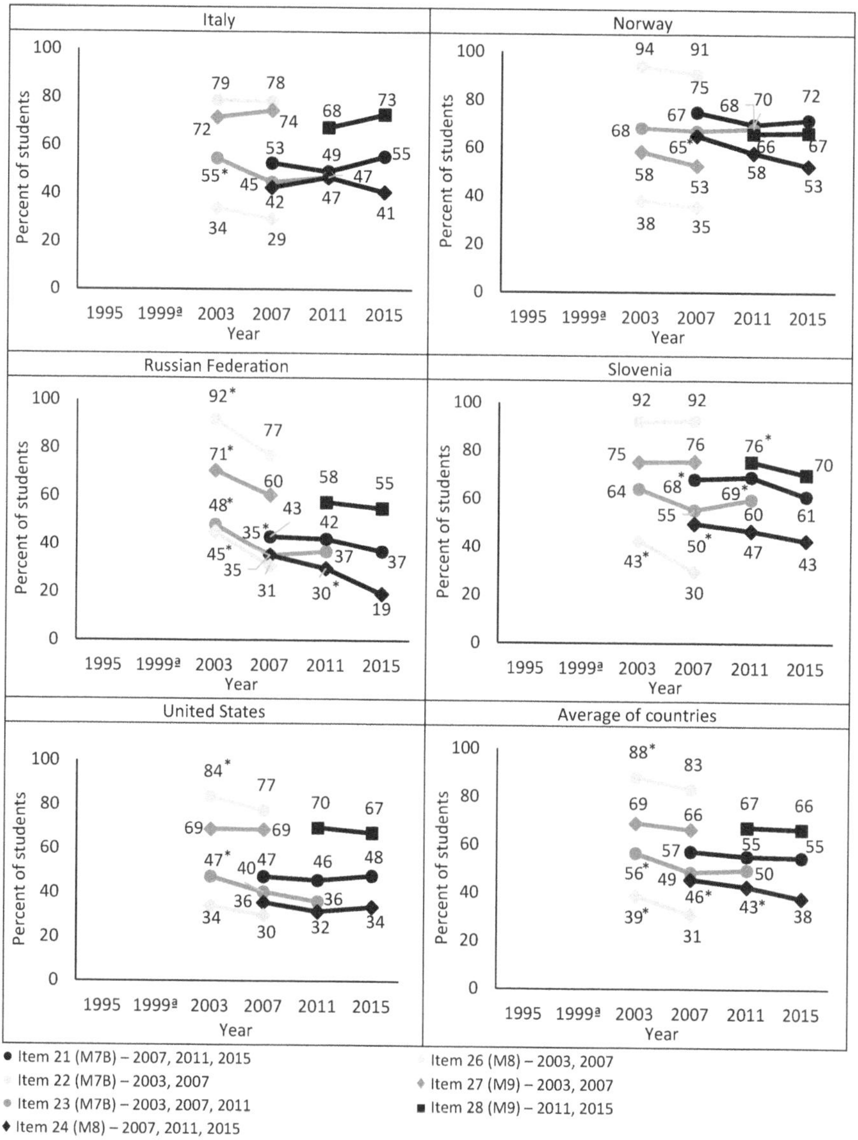

Fig. 4.50 Trends in the percentage of grade four students with errors and misunderstandings about linear equations, 2003–2015. *Notes* Mathematics errors and misunderstandings: M7B = not able to generate a correct verbal description given a specific relationship shown in table form, M8 = not able to identify a correct set of numbers that follow a given relationship/rule, M9 = not able to apply algebraic thinking to solve simple real-life problems involving unknowns. *Significantly different from most recent assessment cycle. [a]TIMSS was not administered in 1999 at grade four

4.3.6 Summary of Mathematics Results

We have reported students' performance on the set of items related to linear equations across countries at each grade level (TIMSS Advanced, grade eight, and grade four; Sect. 4.3.1), patterns in student errors and misunderstandings across countries and grade levels (Sects. 4.3.2 and 4.3.3), gender differences in these errors and misunderstandings (Sect. 4.3.4), and trends over multiple assessment years (Sect. 4.3.5). The frequency of specific types of student errors and misunderstandings at each grade level varied across the five countries included in the study. In each country, and at each grade level, there were some errors and misunderstandings that were demonstrated by at least 50% of the students. There were some gender differences at all three grade levels. Most of the measurable gender differences favored males (i.e., a smaller percentage of males than females demonstrated the error or misunderstanding measured by the item), but there were some that favored females (primarily at grade eight). Performance on trend items administered in multiple assessment years showed that the frequency of certain student errors and misunderstandings changed over time. Performance on grade eight items showed a decrease for some errors and misunderstandings and an increase for others in some countries. Some measurable decreases were also observed at grade four, but, in contrast to grade eight, there were no items that showed an increase in the percentage of students demonstrating the error or misunderstanding.

References

Darling, G. (2012). How does force affect motion? *Science and Children, 50*(2), 50–53.

Sneider, C., & Pulos, S. (1983). Children's cosmographies: Understanding the Earth's shape and gravity. *Science Education, 67*(2), 205–221.

Chapter 5
Conclusions About Using TIMSS and TIMSS Advanced Data to Explore Student Misconceptions, Errors, and Misunderstandings in Physics and Mathematics

Abstract Assessment items from twenty years of TIMSS and TIMSS Advanced assessments enabled the identification of specific types of student misconceptions, errors, and misunderstandings related to two core concepts (gravity in physics and linear equations in mathematics). Results across grade levels, genders, and assessment years for five countries (Italy, Norway, the Russian Federation, Slovenia, and the United States) were compared. In physics, misconceptions and misunderstandings related to gravity were common across all five countries; for most misconceptions at each grade level, at least 25% of students demonstrated the misconception, and, in some countries, >50% of students demonstrated certain misconceptions. Errors and misunderstandings related to linear equations were extremely common across all five countries; on average >50% of students demonstrated errors at each grade level. Gender differences were found at all three grade levels, but to a greater extent in physics than in mathematics. Classroom teachers who are aware of the misconceptions or types of errors students may make will be able to plan for and provide additional support to their students when they are teaching these concepts. TIMSS resources can provide in-depth information about students' level of understanding, and their misconceptions and errors, across a range of core mathematics and science concepts. Access to released assessment items, scoring rationales, and actual student responses may allow researchers to undertake even richer secondary data analysis.

Keywords Errors · Gravity · International large-scale assessment · Linear equations · Mathematics · Misconceptions · Physics · Science · Student achievement · Trend analysis · Trends in International Mathematics and Science Study (TIMSS)

This report illustrates how item-level diagnostic data from TIMSS and TIMSS Advanced can be used to provide in-depth information about students' level of understanding, and specific types of misconceptions, errors, and misunderstandings, related to core physics and mathematics concepts across grade levels (specifically

T. Neidorf et al., *Student Misconceptions and Errors in Physics and Mathematics*, IEA Research for Education 9, https://doi.org/10.1007/978-3-030-30188-0_5

gravity and linear equations in this study). We (1) summarize the results across both physics and mathematics; (2) discuss limitations and further applications of our methodology; (3) consider implications related to instruction in physics and mathematics; and (4) describe some implications for future TIMSS assessment design and reporting.

5.1 Summary of Results Across Physics and Mathematics

The frequency of specific types of student misconceptions, errors, and misunderstandings related to gravity and linear equations at each grade level varied across the five countries included in the study: Italy, Norway, the Russian Federation, Slovenia, and the United States. We compare misconceptions, errors, and misunderstandings for both physics and mathematics by: (1) patterns in misconceptions, errors, and misunderstandings across countries and grade levels; (2) gender differences in misconceptions, errors, and misunderstandings; and (3) trends in misconceptions, errors, and misunderstandings over time (see Tables 4.1 and 4.21 for the specific codes used to refer to misconceptions, errors, and misunderstandings related to gravity and linear equations).[1]

5.1.1 Patterns in Misconceptions, Errors, and Misunderstandings Across Countries and Grades

In analyzing the patterns in student misconceptions, errors, and misunderstandings related to gravity and linear equations (Tables 5.1 and 5.2), we determined the average percentage of students with the misconception, error, or misunderstanding across the corresponding set of items.

In physics (Table 5.1), misconceptions and misunderstandings related to gravity were generally quite common across all five countries. For most misconceptions at each grade level, on average across items, at least 25% of students demonstrated the misconception, and, in some countries, at least 50% of all students demonstrated certain misconceptions. In TIMSS Advanced, misconceptions held by $\geq 50\%$ of students included P1B ("objects thrown upward have no acceleration at their maximum height") in Italy, P2 ("the time on the way up and the time on the way down are not equal") in both Italy and the Russian Federation, and P1C ("gravitational acceleration is always in the direction of motion/velocity") in the United States. At grade four, misconceptions held by $\geq 50\%$ of students included P3B ("gravity alone cannot cause an object initially at rest to start moving") in Italy, and misconception P4C ("gravity can make objects move in other directions that are not

[1]The codes are also defined in the Notes on the tables in Chap. 5.

'down' toward the surface of Earth") in Norway. In contrast, at grade eight there were no misconceptions demonstrated by $\geq 50\%$ of students in any country. There were three misconceptions (one at each grade level) where in all or nearly all countries <25% of students demonstrated the misconception: P1A ("gravitational force (acceleration) acting on objects near Earth's surface is not constant but changes with the height of the object above the surface") in TIMSS Advanced, P4A ("gravitational force causes objects to fall 'down' (in an 'absolute downward' direction in space) rather than toward the center of Earth") at grade eight, and P4B ("gravity pushes upward on objects sitting on a solid surface and on objects that are moving upward") at grade four.

In mathematics (Table 5.2), errors and misunderstandings related to linear equations were extremely common across all five countries; on average, >50% of students showed most types of errors at each grade level. Errors and misunderstandings with lower percentages of students across countries were M3B ("demonstrates confusion between slope and intercept") and M6 ("not able to translate relationships shown in table form into a mathematical equation") at grade eight, and M8 ("not able to identify a correct set of numbers that follow a given relationship/rule") at grade four.

5.1.2 Gender Differences in Misconceptions, Errors, and Misunderstandings

Gender differences in misconceptions, errors, and misunderstandings related to gravity (Table 5.3) and linear equations (Table 5.4) were found at all three grade levels, but to a greater extent in physics than in mathematics. In these summary exhibits, the percentages shown reflect the maximum female–male difference across the items measuring each misconception, error, or misunderstanding.

On average across the five countries, there were gender differences found for all but three gravity misconceptions or misunderstandings: P1C ("gravitational acceleration is always in the direction of motion/velocity") and P3A ("gravity acts only on falling objects, but not on objects at rest or moving upward") in TIMSS Advanced, and P4A ("gravitational force causes objects to fall "down" (in an "absolute downward" direction in space) rather than toward the center of Earth") at grade eight. In comparison, average gender differences were found for about half of the errors or misunderstandings related to linear equations. In physics, there were higher percentages of female students with the misconceptions related to gravity in all countries, with the exception of Italy for misconception P1B ("objects thrown upward have no acceleration at their maximum height where the instantaneous velocity is zero") in TIMSS Advanced, where the percentage of males was higher.

In mathematics, there were five types of errors or misunderstandings related to linear equations with significantly higher percentages of males in at least one country. This applied to four misunderstandings at grade eight: M3B ("demonstrates confusion between slope and intercept of an equation") in Italy and the

Table 5.1 Summary of physics misconceptions and misunderstandings related to gravity across items at each grade level, by country, 1995–2015

Misconceptions and misunderstandings	Items	Percentage of students with misconception or misunderstanding (averaged across items) (%)																	
		TIMSS Advanced						Grade 8						Grade 4					
		ITA	NOR	RUS	SVN	USA	Average of countries	ITA	NOR	RUS	SVN	USA	Average of countries	ITA	NOR	RUS	SVN	USA	Average of countries
Performance objective 1: Determine the acceleration of thrown objects (after they are released)																			
P1A	TIMSS Advanced: Item 1A_V1	16	4	6	7	7	8												
P1B	TIMSS Advanced: Item 1A_V2	71	39	48	44	41	48												
P1C	TIMSS Advanced: Item 2	–	27	34	44	53	39												
Performance objective 2: Determine the time duration between different points on the path of a thrown object																			
P2	TIMSS Advanced: Item 1B	80	37	51	28	36	46												
Performance objective 3: Determine the effect of gravitational force on moving objects or on objects at rest																			
P3A	TIMSS Advanced: Item 3 Grade 8: Items 4, 5, 6, 7, 8	33	22	42	3	–	25	45	43	37	30	31	37						
P3B	Grade 8: Item 9 Grade 4: Items 10, 11							42	32	26	27	22	30	55	41	33	49	33	42
Performance objective 4: Identify the direction of the force due to gravity																			
P4A	Grade 8: Item 12							19	8	13	11	17	14						
P4B	Grade 4: Items 13, 14_V1													25	25	6	11	18	17
P4C	Grade 4: Items 14_V2, 15, 16													38	50	33	46	31	40

Key

- (dark shading) ≥ 50% of students
- (light shading) 25.0–49.9 % of students
- (no shading) < 25% of students

Notes Physics misconceptions and misunderstandings: P1A = gravitational force (acceleration) acting on objects near Earth's surface is not constant but changes with the height of the object above the surface, P1B = objects thrown upward have no acceleration at their maximum height where the instantaneous velocity is zero (the instant it stops moving upward and reverses direction), P1C = gravitational acceleration is always in the direction of motion/velocity (rather than a constant acceleration directed toward the center of Earth), P2 = the time on the way up and the time on the way down are not equal (the downward acceleration due to gravity is not treated as constant), P3A = gravity acts only on falling objects, but not on objects at rest (on the ground or sitting on another surface) or on objects that are moving upward, P3B = gravity alone cannot cause an object initially at rest to start moving; it requires another force/push, P4A = gravitational force causes objects to fall "down" (in an "absolute downward" direction in space) rather than toward the center of Earth, P4B = gravity pushes upward on objects sitting on a solid surface and on objects that are moving upward, P4C = gravity can move objects in other directions that are not "down" toward the surface of Earth. ITA = Italy, NOR = Norway, RUS = Russian Federation, SVN = Slovenia, and USA = United States

– Data not available (see Appendix for country-specific notes)

Table 5.2 Summary of mathematics errors and misunderstandings related to linear equations across items at each grade level, by country, 1995–2015

Errors and misunderstandings	Items	Percentage of students with error or misunderstanding (averaged across items) (%)																	
		TIMSS Advanced						Grade 8						Grade 4					
		ITA	NOR	RUS	SVN	USA	Average of countries	ITA	NOR	RUS	SVN	USA	Average of countries	ITA	NOR	RUS	SVN	USA	Average of countries
Performance objective 1: Interpret the solution to a system of linear equations to answer a question or solve a problem in real life contexts																			
M1	TIMSS Advanced: Item 1A	78	38	60	64	43	57												
Performance objective 2: Solve systems of linear equations in two variables																			
M2A	TIMSS Advanced: Item 1B G8: Item 2	75	57	64	61	60	63	78	76	75	68	61	72						
M2B	Grade 8: Items 3, 4, 5							87	88	60	83	69	77						
Performance objective 3: Interpret the meanings of slope and y-intercept in linear equations or graphs																			
M3A	Grade 8: Item 6							98	89	90	86	69	86						
M3B	Grade 8: Items 7, 8, 9							22	30	12	22	18	19						
Performance objective 4: Relate algebraic equations to their graphical representations (and vice-versa)																			
M4A	Grade 8: Item 10							79	90	52	89	56	73						
M4B	Grade 8: Items 11, 12							65	78	48	63	55	62						
Performance objective 5: Write equations to represent situations																			
M5	Grade 8: Items 13, 14, 15, 16							58	72	36	56	44	48						
Performance objective 6: Given pairs of numbers in tables or ordered pairs, generate an algebraic equation of the relationship between the two variables																			
M6	Grade 8: Items 17, 18, 19							44	43	37	30	31	37						
Performance objective 7: Given pairs of numbers in tables or ordered pairs, generate a verbal description of the relationship																			
M7A	Grade 8: Item 20							48	60	31	47	32	44						
M7B	Grade 4: Items 21, 22, 23													60	77	50	71	54	62
Performance objective 8: Given a verbal description of a relationship between a set of numbers, generate pairs of whole numbers that follow that relationship (rule)																			
M8	Grade 4: Items 24, 25, 26													40	49	26	42	36	39
Performance objective 9: Apply algebraic thinking to solve simple real-life problems involving unknowns																			
M9	Grade 4: Items 27, 28													74	60	58	73	68	66

Key

(dark shading)	≥ 50% of students
(light shading)	25.0–49.9% of students
(no shading)	< 25%

Notes Mathematics errors and misunderstandings: M1 = not able to use slope and intercept to provide an argument in support of the solution to a real-life problem situation, M2A = not able to apply the procedure correctly to solve a real-life problem situation, M2B = not able to apply the procedure correctly to solve non-contextualized problems, M3A = not able to relate slope with steepness of a line, M3B = demonstrates confusion between slope and intercept of an equation, M4A = not able to correctly identify the graph of an equation, M4B = not able to translate graphical representations into a mathematical equation or verbal description of a linear relationship, M5 = not able to translate verbal descriptions into a correct mathematical equation, M6 = not able to translate relationship shown in table form into a mathematical equation, M7A = not able to generate a correct verbal description given a specific relationship in the form of ordered pairs, M7B = not able to generate a correct verbal description given a specific relationship shown in table form, M8 = not able to identify a correct set of numbers that follow a given relationship/rule, M9 = not able to apply algebraic thinking to solve simple real-life problems involving unknowns. ITA = Italy, NOR = Norway, RUS = Russian Federation, SVN = Slovenia, and USA = United States

Table 5.3 Summary of gender differences in physics misconceptions and misunderstandings related to gravity across items at each grade level, by country, 1995–2015

Misconceptions and misunderstandings	Items	Percentage of students with misconception or misunderstanding (maximum female–male difference across items) (%)																	
		TIMSS Advanced						Grade 8						Grade 4					
		ITA	NOR	RUS	SVN	USA	Average of countries	ITA	NOR	RUS	SVN	USA	Average of countries	ITA	NOR	RUS	SVN	USA	Average of countries
Performance objective 1: Determine the acceleration of thrown objects (after they are released)																			
P1A	TIMSS Advanced: Item 1A_V1	9	.	6	.	10	5												
P1B	TIMSS Advanced: Item 1A_V2	-9	20	.	17	16	9												
P1C	TIMSS Advanced: Item 2	–	.	.	.	.	.												
Performance objective 2: Determine the time duration between different points on the path of a thrown object																			
P2	TIMSS Advanced: Item 1B	13	15	16	.	14	13												
Performance objective 3: Determine the effect of gravitational force on moving objects or on objects at rest																			
P3A	TIMSS Advanced: Item 3 Grade 8: Items 4, 5, 6, 7, 8	.	.	.	.	–	.	21	.	13	.	8	7						
P3B	Grade 8: Item 9 Grade 4: Items 10, 11							.	.	.	9	.	4	11	11	.	.	11	7
Performance objective 4: Identify the direction of the force due to gravity																			
P4A	Grade 8: Item 12							.	.	.	.	.	.						
P4B	Grade 4: Items 13, 14_V1													.	.	.	.	8	3
P4C	Grade 4: Items 14_V2, 15, 16													.	18	9	17	8	11

Key

- (dark grey) Higher percentage of female students with the misconception or misunderstanding
- (light grey) Higher percentage of male students with the misconception or misunderstanding
- . No statistically significant gender difference

Notes Physics misconceptions and misunderstandings: P1A = gravitational force (acceleration) acting on objects near Earth's surface is not constant but changes with the height of the object above the surface, P1B = objects thrown upward have no acceleration at their maximum height where the instantaneous velocity is zero (the instant it stops moving upward and reverses direction), P1C = gravitational acceleration is always in the direction of motion/velocity (rather than a constant acceleration directed toward the center of Earth), P2 = the time on the way up and the time on the way down are not equal (the downward acceleration due to gravity is not treated as constant), P3A = gravity acts only on falling objects, but not on objects at rest (on the ground or sitting on another surface) or on objects that are moving upward, P3B = gravity alone cannot cause an object initially at rest to start moving; it requires another force/push, P4A = gravitational force causes objects to fall "down" (in an "absolute downward" direction in space) rather than toward the center of Earth, P4B = gravity pushes upward on objects sitting on a solid surface and on objects that are moving upward, P4C = gravity can move objects in other directions that are not "down" toward the surface of Earth. ITA = Italy, NOR = Norway, RUS = Russian Federation, SVN = Slovenia, and USA = United States

– Data not available (see Appendix for country-specific notes)

Table 5.4 Summary of gender differences in mathematics errors and misunderstandings related to linear equations across items at each grade level, by country, 1995–2015

Errors and misunderstandings	Items	Percentage of students with error or misunderstanding (averaged across items) (%)																	
		TIMSS Advanced						Grade 8						Grade 4					
		ITA	NOR	RUS	SVN	USA	Average of countries	ITA	NOR	RUS	SVN	USA	Average of countries	ITA	NOR	RUS	SVN	USA	Average of countries
Performance objective 1: Interpret the solution to a system of linear equations to answer a question or solve a problem in real life contexts																			
M1	TIMSS Advanced: Item 1A	12	.	15	18	.	8												
Performance objective 2: Solve systems of linear equations in two variables																			
M2A	TIMSS Advanced: Item 1B G8: Item 2	9	.	12	19	.	7	.	.	.	.	.	.						
M2B	Grade 8: Items 3, 4, 5							.	.	.	5	.	4						
Performance objective 3: Interpret the meanings of slope and y-intercept in linear equations or graphs																			
M3A	Grade 8: Item 6							.	.	6	.	.	.						
M3B	Grade 8: Items 7, 8, 9							-5	.	.	.	-5	-4						
Performance objective 4: Relate algebraic equations to their graphical representations (and vice-versa)																			
M4A	Grade 8: Item 10							.	.	.	6	.	.						
M4B	Grade 8: Items 11, 12							.	10	11	.	.	4						
Performance objective 5: Write equations to represent situations																			
M5	Grade 8: Items 13, 14, 15, 16							.	.	−11	.	-6	.						
Performance objective 6: Given pairs of numbers in tables or ordered pairs, generate an algebraic equation of the relationship between the two variables																			
M6	Grade 8: Items 17, 18, 19							11	.	−10	.	.	.						
Performance objective 7: Given pairs of numbers in tables or ordered pairs, generate a verbal description of the relationship																			
M7A	Grade 8: Item 20							.	.	.	−10	.	.						
M7B	Grade 4: Items 21, 22, 23													.	11	.	.	.	.
Performance objective 8: Given a verbal description of a relationship between a set of numbers, generate pairs of whole numbers that follow that relationship (rule)																			
M8	Grade 4: Items 24, 25, 26													.	15	.	.	−7	.
Performance objective 9: Apply algebraic thinking to solve simple real-life problems involving unknowns																			
M9	Grade 4: Items 27, 28													.	.	.	9	.	5

Key

(dark shading) Higher percentage of female students with the error or misunderstanding

(light shading) Higher percentage of male students with the error or misunderstanding

. No statistically significant gender difference

Notes Mathematics errors and misunderstandings: M1 = not able to use slope and intercept to provide an argument in support of the solution to a real-life problem situation, M2A = not able to apply the procedure correctly to solve a real-life problem situation, M2B = not able to apply the procedure correctly to solve non-contextualized problems, M3A = not able to relate slope with steepness of a line, M3B = demonstrates confusion between slope and intercept of an equation, M4A = not able to correctly identify the graph of an equation, M4B = not able to translate graphical representations into a mathematical equation or verbal description of a linear relationship, M5 = not able to translate verbal descriptions into a correct mathematical equation, M6 = not able to translate relationship shown in table form into a mathematical equation, M7A = not able to generate a correct verbal description given a specific relationship in the form of ordered pairs, M7B = not able to generate a correct verbal description given a specific relationship shown in table form, M8 = not able to identify a correct set of numbers that follow a given relationship/rule, M9 = not able to apply algebraic thinking to solve simple real-life problems involving unknowns. ITA = Italy, NOR = Norway, RUS = Russian Federation, SVN = Slovenia, and USA = United States

United States, M5 (“not able to translate verbal descriptions into a correct mathematical equation”) in the Russian Federation and the United States, M6 (“not able to translate relationship shown in table form into a mathematical equation”) in the Russian Federation, and M7A (“not able to generate a correct verbal description given a specific relationship in the form of ordered pairs”) in Slovenia. There was also one misunderstanding at grade four (M8, “not able to identify a correct set of numbers that follow a given relationship/rule”) in the United States.

These results, based on item-level data, track what was found in scale score gender differences in the international reports from TIMSS and TIMSS Advanced in 2015 (Martin et al. 2016; Mullis et al. 2016a, b). In the five countries included in our study, males generally outperformed females in the relevant science content domains that covered the gravity topic in physics (mechanics and thermodynamics in TIMSS Advanced, physics at grade eight, and physical science at grade four). In contrast, there were fewer and smaller gender differences in the mathematics content domains that covered linear equations (algebra in TIMSS Advanced and grade eight, and number at grade four), and not all of these favored males. At grade eight, females scored higher than males for algebra in Italy, Slovenia, and the United States. Both in the item-level percentage of students with misconceptions, errors, or misunderstandings in this report, and in the subscale scores in the international reports, gender differences in both physics and mathematics were generally higher in TIMSS Advanced than at the lower grade levels. However, there were differences between physics and mathematics in the patterns of gender differences across grades in each country.

5.1.3 Trends in Patterns of Misconceptions, Errors, and Misunderstandings Over Time

The trend patterns across both physics (Figs. 4.23 and 4.24) and mathematics (Figs. 4.49 and 4.50) indicate some interesting differences over the assessment years in the frequency of misconceptions, errors, and misunderstandings demonstrated on items related to gravity and linear equations across countries and grade levels.[2]

Italy

There were very few measurable differences in the percentage of students with misconceptions, errors, and misunderstandings over time. Significant differences were found in mathematics for one item at grade four, where the frequency of misunderstanding M7B decreased between 2003 and 2015, and for one item at grade eight, where the frequency of misconception M3A increased slightly between 2011 and 2015. In physics, misconception P3A decreased in frequency between 2011 and 2015.

[2]The definitions for each misconception, error, or misunderstanding code referred to in this section can be found in Chap. 4 (Tables 4.1 and 4.21) and also in the notes in Tables 5.1 and 5.2.

Norway

There were no changes in the frequency of gravity misconceptions or misunderstandings at either grade four or grade eight. In mathematics, there was one item at grade four that showed a decrease in the frequency of misunderstanding M8 from 2007 to 2015. In contrast, there were two mathematics items at grade eight where the frequency increased for the types of errors and misunderstandings: in M4B (between 2007 and 2015) and M7A (between 1995 and 2003), and one item measuring misunderstanding M3A, where the frequency decreased between 2011 and 2015.

Russian Federation

Across grades and subjects, the greatest number of items showing trend differences was in the Russian Federation (10 items total). Most of the trend differences were in grade four, where the percentage of students with misconceptions related to gravity (three of three items, measuring misconceptions P3B and P4B) and misunderstandings related to linear equations (five of seven items, measuring misunderstandings M7B, M8 and M9) decreased over time. In physics, the frequency of misconception P3B also decreased at grade eight. The only case of an increase occurred in grade eight mathematics (misunderstanding M3A).

Slovenia

The number of items with trend differences were greater in grade eight than grade four for physics (two versus three items) and greater in grade four than grade eight (four versus three items) for mathematics. At grade eight, there were some significant decreases over time in the frequency of misconceptions, errors, or misunderstandings, and errors related to gravity (P3A and P3B) and linear equations (M3A and M5). In mathematics, however, there was an increase in misunderstanding M7A at grade eight between 1995 and 2003. At grade four, the frequency of misconceptions and misunderstandings decreased on one item in physics (measuring misconception P3B) and on four items in mathematics (measuring misunderstandings M7B, M8, and M9).

United States

The number of items with trend differences in the United States was the same as in the Russian Federation (10 items total in both countries). In contrast to the Russian Federation, however, the majority of trend differences in the United States were in grade eight mathematics, where the frequency decreased across assessment years on five items measuring errors or misunderstandings for M2B, M3A, M4A, M6, and M7A, and increased on one item (misunderstanding M4B). In physics, the frequency of misconceptions related to gravity were found to decrease on one item at grade eight (misconception P3A) and one item at grade four (misconception P3B).

5.2 Limitations and Further Applications of the Methodology

For our study, we used item-level data from the TIMSS international database (https://www.iea.nl/data) and, therefore, we were limited by the specific types of diagnostic data provided. In large-scale assessments like TIMSS, there is always a balance between the resources required for scoring, maintaining high reliability of scoring, and collecting diagnostic data that will provide information for tracking specific types of misconceptions, errors, and misunderstandings. Generally, for mathematics items there is a correct response and an incorrect response, with only a few items in the set that we used for our study being scored using a two-point scoring guide, where one point was given for a partial response. Similarly, there were only a few CR items worth one score point that used diagnostic scoring guides to track specific types of incorrect responses. In the case of physics items, there were slightly more CR items that used diagnostic scoring guides to track particular types of incorrect responses. For future studies similar to ours, more items with scoring codes that track the different types of errors that students make would be useful, particularly in mathematics.

The information produced by MC items is also limited by the guessing factor involved for such items. In general, for the same misconception, error, or misunderstanding, the percentage of students demonstrating the misconception or error may be higher for CR items than for MC items. The information provided by MC items could be enhanced if the distractors tracked important types of conceptual misunderstanding rather than the computational errors that students can make while solving the problem.

For the CR items, unless there was a specific diagnostic code to track particular misconceptions, errors, or misunderstandings, the reporting of more general misunderstandings and errors included all incorrect responses (including blanks). In doing this, we assumed that students who left the item blank did not know how to apply the concept or mathematical procedure in order to solve the problem, similar to other incorrect responses where students do not make an attempt at the item (e.g., random marks or off-task comments). However, it is difficult to know why students did not answer the item. Therefore, the percentage of students with misunderstandings or errors on these types of items may be inflated.

The TIMSS and TIMSS Advanced assessments are designed to provide reliable overall scores for science (or physics in TIMSS Advanced) and mathematics, and for each content domain. However, the sample sizes for the item-level statistics used in this report (percent correct and percentage of students demonstrating different types of misconceptions, errors, and misunderstandings) are relatively low.[3] As a result, many of the observed differences across countries, genders, and assessment years were not statistically significant. Also, as result of the booklet

[3]The number of students responding to each item is shown in the supplementary tables provided at www.iea.nl/publications/RfEVol9.

rotation scheme used in the TIMSS assessment design, only about one in every seven students get the same item; for TIMSS Advanced, about one in every three students get the same item.[4] This means a very small number of students in each class take the same item, which particularly affects the ability to report gender differences within countries.

To generalize beyond students' performance on individual items, a larger set of items that measure each type of misconception, error, or misunderstanding would be needed in each assessment cycle. In that case, "misconception indices," based on the average percentage of students with misconceptions across items, could be computed and tested for reliability in order to compare the frequency of these misconceptions on a broader range of items across countries and grade levels.

In addition, it would be interesting to follow a cohort of students to track the percentage of students with particular misconceptions, errors, and misunderstandings over time (e.g., students who were grade four in 2007, grade eight in 2011, and then, TIMSS Advanced in 2015). This would provide international data for understanding how students conceptualize a topic of interest as they progress through the grades and how similar or different the patterns in misconceptions, errors, and misunderstandings are across countries. Again, more items related to the topic of interest would be needed in each assessment cycle for a reliable measure.

While this report focused on specific types of misconceptions, errors, and misunderstandings related to the topics of gravity in physics and linear equations in mathematics, the general methodology that we describe can be applied to a range of science and mathematics topics covered in TIMSS and TIMSS Advanced to trace misconceptions, errors, and misunderstandings across two or three grade levels and better understand students' performance on those topics in science and mathematics. Another area that countries could continue to explore is the pattern of misconceptions, errors, and misunderstandings at one grade only, as was done in the United States for TIMSS Advanced (Provasnik et al. 2019). This could produce rich information about the misconceptions, errors, and misunderstandings that students at a specific grade have across different content domains.

We examined differences in misconceptions, errors, and misunderstandings by gender, but there are many other demographic variables available in TIMSS and TIMSS Advanced that could be analyzed. Countries could also look at differences by region, school type, or course type, as was done in the TIMSS Advanced report for the United States (Provasnik et al. 2019).

A better understanding of the misconceptions, errors, and misunderstandings over the assessment years could be achieved by investigating what is happening at the country level in the education system. A change in the curriculum, a change in the approach to teaching, or a change in the emphasis on the various types of learning strategies that could have resulted in a change in the pattern of

---

[4]For additional information on the assessment design for TIMSS 2015 and TIMSS Advanced 2015, please see https://timssandpirls.bc.edu/timss2015/downloads/T15_FW_Chap4.pdf and https://timssandpirls.bc.edu/timss2015-advanced/downloads/TA15_FW_Chap4.pdf, respectively.

misconceptions, errors, and misunderstandings made by students in different assessment years, merits further investigation. This kind of information, along with the methodology that we used for this report, could support teachers' and educators' efforts to improve instruction in the classroom. While it is beyond the scope of this report to explore curricular changes in the five different countries included in our study, further research could focus on this aspect. The TIMSS and TIMSS Advanced encyclopedias, teacher questionnaires, and country-level curriculum questionnaires, and results from the test curriculum matching analyses provide context for results from this type of study in terms of possible changes in policy, curriculum, or instruction across assessment cycles or grades (Martin et al. 2016; Mullis et al. 2016a, b, c, d). It should be noted, however, that any future research connecting curriculum changes to patterns and trends in the specific types of misconceptions, errors, and misunderstandings discussed in this report would likely require a more detailed analysis of curriculum documents from each country.

5.3 Implications Related to Instruction

In this report, we have discussed different types of misconceptions, errors, and misunderstandings related to gravity and linear equations that were demonstrated by TIMSS Advanced students in their final year of secondary school, and showed how these were connected to related misconceptions and a lack of foundational understanding about these concepts at grades four and eight. By identifying specific misconceptions, errors, and misunderstandings related to these core concepts, the findings from this type of study support the teaching, learning, and reinforcement of core concepts throughout school. Classroom teachers who are aware of the misconceptions or types of errors that students may make will be able to plan for and provide additional support to their students when they are teaching these concepts. Using released TIMSS and TIMSS Advanced items as additional resources may enable science and mathematics educators to identify misconceptions, develop pre-assessments, and provide focused instruction for their students.

In physics, our study showed that many TIMSS Advanced students still have difficulty understanding the effects of constant acceleration due to gravity on motion. The types of misconceptions related to gravity (and to forces and motion in general) described in previous smaller-scale studies across different grade levels were found to persist in the nationally representative TIMSS and TIMSS Advanced samples, including TIMSS Advanced students who had taken more advanced coursework in physics. In particular, it is of concern that many students in TIMSS Advanced across countries did not grasp the concept that the force (acceleration) due to gravity is a constant for thrown objects, instead indicating there was no acceleration at the maximum height and that acceleration was always in the direction of motion/velocity, rather than a constant acceleration directed toward the center of Earth. The misconception held by TIMSS Advanced students that acceleration due to gravity is not constant may arise from related misconceptions about the force of gravity at earlier grades.

The TIMSS data revealed that a lack of basic understanding of gravitational force at the lower grades can lead to misconceptions at higher grade levels, including the misconceptions that gravity acts only on falling objects, that gravity alone cannot cause an object initially at rest to start moving without another force/ push, and that the force due to gravity is directed upward for an object at rest sitting on a surface or for objects that are moving upward.

Based on the types of gravity misconceptions found across grade levels, it is important for teachers at all grades to expose their students to a broad range of problem-solving contexts that will develop and evaluate their ability to apply their understanding of the concepts related to force and motion. In addition, pre-assessments and hands-on activities have been found to be important in identifying and addressing student misconceptions and developing their knowledge of forces (Darling 2012).

In mathematics, our report showed various conceptual stages where students have problems or make errors on the items involving linear equations that have been discussed in previous studies (Simon and Blume 1994; Stump 2001; Kalchman and Koedinger 2005; Caglayan and Olive 2010). These are the areas where focused instruction is needed for students to make the leap toward being well-versed in that concept. For example, one of the findings was that a higher percentage of students at grade eight were able to translate a graphical representation into a verbal description as opposed to an algebraic equation. This could mean that students are able to understand the relationship represented by the graph of a line, but they are not well-versed in the symbolic representation of a line, what each symbol means, and how they are related. Instruction needs to focus on these aspects, with an emphasis on understanding that goes beyond using equations to find the value of one variable when the other is given.

Similarly, students at each grade level find solving real-life problems more difficult than solving non-contextualized mathematics problems (item 1 in TIMSS Advanced, item 15 at grade eight, and items 24 and 25 at grade four). Students have difficulty solving real-life problems that require reading the context, understanding it, and then translating the problem into mathematics language to find what they need to do to solve the problem. Instruction across the grade levels needs to include more and different types of application problems that go beyond pure computation.

5.4 Implications for Future TIMSS Assessment Design and Reporting

While TIMSS is designed primarily to monitor system-level achievement trends in a global context, another important outcome of the study is the diagnosis of common learning difficulties in mathematics and science, as evidenced by misconceptions and errors (Mullis and Martin 2013a). Thus, TIMSS items and associated scoring guides are developed to allow identification of widespread student misunderstandings that, in turn, could lead to curricular or instructional improvements (Mullis and Martin 2013b). For example, TIMSS MC items use plausible distracters that are based on likely student errors or misconceptions.

CR items are scored using the TIMSS two-digit diagnostic scoring system, which allowed us to classify responses based on the method used in solving a problem, and track common errors or misconceptions. However, because scoring of CR items is a significant cost factor for the TIMSS countries, diagnostic scoring codes for specific response types are developed parsimoniously, such that only the codes with apparent value for educational improvement are included in the scoring guides (Mullis and Martin 2013b). As a result, the TIMSS item-level diagnostic data are limited to pre-defined distractors and diagnostic codes included to capture only the predominant correct and incorrect approaches/strategies used by students across all participating countries.

Despite this design restriction, our report demonstrated that access to specific TIMSS resources, namely released assessment items, CR item scoring guides, and item-level diagnostic data, can provide in-depth information about students' level of understanding and their misconceptions and errors across a range of core mathematics and science concepts. In addition to these critical TIMSS resources, future cycles of TIMSS may consider offering two additional resources: access to more complete scoring rationales for both CR and MC items, and actual student responses. Such resources would allow even richer secondary data analysis of mathematics and science concepts, and misconceptions, errors, and misunderstandings.

TIMSS items and scoring guides are developed with great care and thoughtfulness, with specific reasons for including each MC distractor item and each response code for the scoring guides of the CR items. Researchers would benefit greatly from having access to the rationales for the inclusion of specific distractors and specific response codes in TIMSS items.

Access to scoring rationales can be coupled with the potential benefits of eTIMSS, an electronic version of TIMSS. The 2019 administration of TIMSS begins the transition to administering the assessments in the eTIMSS digital format, allowing enhanced assessment of complex areas of the TIMSS framework that are difficult to measure with the paper-and-pencil format. In addition, eTIMSS will be able to capture students' actual responses to items in an easily accessible digital format. Traditionally, TIMSS provides access to achievement data files containing the actual responses to the MC items and the codes assigned to the CR items through the TIMSS scoring guides. Starting with the 2019 cycle, eTIMSS has the potential to provide access to a new international data file for students' responses that are captured via keyboard/number pad input. This new TIMSS resource has high value for researchers, since it potentially provides even deeper insights into what students know and are able to do, including common misconceptions, errors, and misunderstandings.

As discussed in Sect. 5.2, a more focused effort on providing diagnostic outcomes from TIMSS would require the inclusion of a larger number of items at each grade level that measure certain core concepts and misconceptions of interest. Also, sets of items related to a particular concept would need to be kept secure and administered in multiple assessments in order to track trends in students' understanding and how their misconceptions about concepts develop or vary over time.

The TIMSS and TIMSS Advanced assessments cover the framework objectives in each content domain with enough items to permit subscale reporting. However, each individual topic is measured by a small number of items distributed across the assessment booklets. Since each booklet includes only a portion of the total item pool, only a small subset of students in each country are likely to take items related to a particular topic. Therefore, while scores are provided at the content domain level, it is not possible to obtain reliable student-level data on a set of items that measure a particular topic within a content domain. To provide the best diagnostic information, students would have to take multiple items related to a specific topic in a single assessment (not possible with the current assessment design) in order to generalize beyond performance on individual items. One possible way to accomplish this would be to select one topic to explore in more depth and develop a block of 10–15 items that measure particular types of misconceptions, errors, and misunderstandings related to this topic. These special item blocks would be administered to a subset of students in the national samples, providing enough student-level data to support diagnostic reporting of the selected topic.

As also discussed in Sect. 5.2, it would be interesting to follow the same cohort of students across grade levels to track how their conceptual understanding of a concept develops with schooling over the years. TIMSS has a "quasi-longitudinal" design that permits this type of study, with the grade four and grade eight assessments being conducted every four years (see https://www.iea.nl/timss). However, in order to track the patterns of misconceptions, errors, and misunderstandings across grade levels, a change would be needed in the assessment design to include a block of cross-grade items (or a related block of items at each grade level) that measure a particular topic in consecutive assessment cycles. TIMSS Advanced has been administered less often than TIMSS,[5] so measuring the same cohort of students from grade four to the final year of secondary school would require putting TIMSS and TIMSS Advanced on the same assessment schedule. Even if a cohort is not tracked across all three grade levels though, monitoring the frequency of

[5]The 2015 assessment year was unusual in that all three assessments were administered, and there are data available for the same cohort of students (e.g., 2007 grade 4, 2011 grade 8, and 2015 TIMSS Advanced).

misconceptions, errors, and misunderstandings related to one topic of interest between grade four and grade eight could be a useful addition for future TIMSS cycles.

References

Caglayan, G., & Olive, J. (2010). Eighth grade students' representations of linear equations based on a cups and tiles model. *Educational Studies in Mathematics, 74*(2), 143–162.

Darling, G. (2012). How does force affect motion? *Science and Children, 50*(2), 50–53.

Kalchman, M., & Koedinger, K. R. (2005). Teaching and learning functions. In M. S. Donovan, & J. D. Bransford (Eds.), *How students learn: History, mathematics, and science in the classroom* (pp. 351–393). Washington, DC: National Academies Press.

Martin, M. O., Mullis, I. V. S., Foy, P., & Hooper, M. (2016). *TIMSS 2015 international results in science*. Chestnut Hill, MA: TIMSS & PIRLS International Study Center, Boston College. Retrieved from: http://timssandpirls.bc.edu/timss2015/international-results/.

Mullis, I. V. S., & Martin, M. O. (Eds.). (2013a). *TIMSS 2015 assessment frameworks*. Chestnut Hill, MA: TIMSS & PIRLS International Study Center, Boston College. Retrieved from http://timssandpirls.bc.edu/timss2015/frameworks.html.

Mullis, I. V. S., & Martin, M. O. (2013b). *TIMSS 2015 item writing guidelines*. Chestnut Hill, MA: TIMSS & PIRLS International Study Center, Boston College. Retrieved from https://timssandpirls.bc.edu/publications/timss/T15_item_writing_guidelines.pdf.

Mullis, I. V. S., Martin, M. O., Foy, P., & Hooper, M. (2016a). *TIMSS 2015 international results in mathematics*. Chestnut Hill, MA: TIMSS & PIRLS International Study Center, Boston College. Retrieved from: http://timssandpirls.bc.edu/timss2015/international-results/.

Mullis, I. V. S., Martin, M. O., Foy, P., & Hooper, M. (2016b). *TIMSS Advanced 2015 international results in advanced mathematics and physics*. Chestnut Hill, MA: TIMSS & PIRLS International Study Center, Boston College. Retrieved from: http://timssandpirls.bc.edu/timss2015/international-results/advanced/.

Mullis, I. V. S., Martin, M. O., Goh, S., & Cotter, K. (Eds.) (2016c). *TIMSS 2015 encyclopedia: Education policy and curriculum in mathematics and science*. Chestnut Hill, MA: TIMSS & PIRLS International Study Center, Boston College. Retrieved from: http://timssandpirls.bc.edu/timss2015/encyclopedia/.

Mullis, I. V. S., Martin, M. O., & Loveless, T. (2016d). *20 years of TIMSS: International trends in mathematics and science achievement, curriculum, and instruction* (pp. 23–37). Chestnut Hill, MA: TIMSS & PIRLS International Study Center, Boston College. Retrieved from: http://timss2015.org/timss2015/wp-content/uploads/2016/T15-20-years-of-TIMSS.pdf.

Provasnik, S., Malley, L., Neidorf, T., Arora, A., Stephens, M., Balestreri, K., Perkins, R., & Tang, J. H. (2019, in press). *U.S. performance on the 2015 TIMSS Advanced mathematics and physics assessments: A closer look* (NCES 2017-020). Washington, DC: US Department of Education, National Center for Education Statistics.

Simon, M. A., & Blume, G. W. (1994). Mathematical modeling as a component of understanding ratio-as-measure: A study of prospective elementary teachers. *Journal of Mathematical Behavior, 13,* 183–197.

Stump, S. L. (2001). High school precalculus students' understanding of slope as measure. *School Science and Mathematics, 101*(2), 81–89.

Appendix
Technical Documentation and Syntax

A.1 Data Considerations for the Study

This section provides additional information about the TIMSS and TIMSS Advanced populations across the assessment years and provides notes on data considerations that are applicable to the items, countries, and assessment years included in this study.

A.1.1 Considerations for 1995 Data

In 1995, TIMSS defined two populations, one in elementary school (population 1) and one in middle school (population 2). Each population included all students enrolled in two adjacent grade levels. Population 1 included the two grades that contained the largest proportion of nine-year-olds (grades three and four in most countries) at the time of testing, and population 2 included the two grades that contained the largest proportion of 13-year-olds (grades seven and eight in most countries) at the time of testing. After the first assessment in 1995, TIMSS assessed only the upper of the two grade levels in elementary school and in middle school (referred to as the TIMSS fourth- and eighth-grade assessments). In this study, data from 1995 include only the upper grades (four and eight), which are comparable to the target populations in all subsequent administrations of TIMSS. Slovenia was the one exception, as described below.

T. Neidorf et al., *Student Misconceptions and Errors in Physics and Mathematics*, IEA Research for Education 9, https://doi.org/10.1007/978-3-030-30188-0

Italy

(1) Grade four. Although Italy participated in the 1995 TIMSS grade four assessment, their data are not comparable for measuring trends to 2015. As a result, their 1995 data were excluded from all analyses.
(2) Grade eight. Although Italy participated in the 1995 TIMSS grade eight assessment, their data are not comparable for measuring trends to 2015. As a result, their 1995 data were excluded from all analyses.
(3) TIMSS Advanced. Italy did not participate in the 1995 TIMSS Advanced assessment.

Slovenia

(1) Grade four. In 1995, Slovenia participated in TIMSS at grades three and four. The 1995 grade three population (comparable to grade four in other countries) is equivalent to Slovenia's target population in subsequent administrations of TIMSS; therefore, the grade three data from 1995 were used for this study.
(2) Grade eight. In 1995, Slovenia participated in TIMSS at grades seven and eight. The 1995 grade seven population (comparable to grade eight in other countries) is equivalent to Slovenia's target population in subsequent administrations of TIMSS; therefore, the grade seven data from 1995 were used for this study.

United States

TIMSS Advanced. The United States changed its sampling procedures from 1995 to 2015. To make the 1995 sample comparable to the 2015 sample, a United States specific COMPARISON variable was added to the 1995 data file to create a subset sample, which was comparable to the 2015 sample. To replicate the United States TIMSS Advanced 1995 results in this report, users need to use the United States' 1995 TIMSS Advanced data files available from the United States' National Center for Education Statistics (NCES) website (https://nces.ed.gov/pubsearch/pubsinfo.asp?pubid=2018127). More details about the changes in the United States sample can be found in chapter 5.11.7 of Averett et al. (2017).

A.1.2 Considerations for 1999 Data

Norway

Grade eight. Norway did not participate in the 1999 TIMSS assessment.

Slovenia

Grade eight. Although Slovenia participated in the 1999 TIMSS grade eight assessment, their data are not comparable for measuring trends to 2015. As a result, their 1999 data were excluded from all analyses.

A.1.3 Considerations for 2008 Data

United States

TIMSS Advanced. The United States did not participate in the 2008 TIMSS Advanced assessment.

A.1.4 Considerations for 2015 Data

Norway

(1) Grade four. In 2015, Norway chose to assess students in grade five to obtain better comparisons with Sweden and Finland. In previous TIMSS cycles, Norway assessed grade four, which is similar to grade three in many other countries because grade one in Norway is considered the equivalent of a year of kindergarten/preschool. To maintain trend with previous TIMSS cycles, in 2015 Norway also collected benchmark data at grade four. This study uses only the grade four data for Norway.

(2) Grade eight. In 2015, Norway chose to assess students in grade nine to obtain better comparisons with Sweden and Finland. In previous TIMSS cycles, Norway assessed grade eight, which is similar to grade seven in many other countries because grade one in Norway is considered the equivalent of a year of kindergarten. To maintain trend with previous TIMSS cycles, in 2015 Norway also collected benchmark data at grade eight. This study uses only the grade eight data for Norway.

Russian Federation

TIMSS Advanced. In 2015, the Russian Federation expanded its advanced mathematics sample to include all students who are currently taking (or have previously taken) advanced courses in mathematics. In previous cycles (1995 and 2008), the Russian Federation sampled only students with six or more hours of mathematics courses per week (defined as the "intensive" stream). In 2015, the Russian Federation sampled students from both the "intensive" stream (six or more hours) and the "profile" stream (4.5 hours); the "intensive" stream data can be analyzed separately for reporting trends relative to previous assessment cycles. For this study, the 2015 data for the Russian Federation reflect the full TIMSS Advanced sample (including students in both the "intensive" and "profile" streams), which means the 2015 population is similar to the TIMSS Advanced populations in other countries.

A.2 Value Labels and Misconception/Error/Misunderstanding Codes Used for Analyses of Physics and Mathematics Items

Tables A.1 and A.2 provide the value labels and misconception/error/misunderstanding codes used for the analyses of the physics and mathematics items.

Table A.1 Value labels and misconception/misunderstanding codes for TIMSS and TIMSS Advanced physics items related to gravity

Physics item number	Permanent identification code in TIMSS	Correct answer	Value labels	Misconception/misunderstanding codes
1A_V1	PA33061A	D	A, B, C, D, 6, 9	B, C
1A_V2	PA33061A	D	A, B, C, D, 6, 9	A
1B	PA33061B	10	10, 79, 96, 99	79, 99
2	PA13063	10	10, 70, 71, 72, 73, 74, 79, 90, 96, 97, 99	70, 71, 72
3	PA23014	10	10, 70, 71, 79, 96, 97, 99	71, 79, 99
4	S032141	D	A, B, C, D, 6, 7, 9	A, B
5	S022012	D	A, B, C, D, 6, 7, 9	A
6	S012075	D	A, B, C, D, 6, 7, 9	A, B
7	S042211	10, 11, 12, 19	10, 11, 12, 19, 70, 71, 79, 96, 99	70, 71, 79, 99
8	S032281	A	A, B, C, D, 6, 7, 9	C
9	S042293A	10	10, 79, 96, 99	79, 99
10	S031311	B	A, B, C, D, 6, 7, 9	D
11	S051147	10	10, 79, 96, 99	79, 99
12	S032714	D	A, B, C, D, 6, 7, 9	A
13	S041119	C	A, B, C, D, 6, 9	A
14_V1	S061048	A	A, B, C, D, 6, 9	C
14_V2	S061048	A	A, B, C, D, 6, 9	B, D
15	S041308	D	A, B, C, D, 6, 9	A, B, C, 9
16	S031313	B	A, B, C, D, 6, 7, 9	A, C

Table A.2 Values and error/misunderstanding codes for TIMSS and TIMSS Advanced mathematics items related to linear equations

Mathematics item number	Permanent ID	Key	Value labels	Error/misunderstanding codes
1A	MA33240	20, 10	20, 10, 11, 79, 96, 99	11, 79, 99
1B	MA33240	20, 11	20, 10, 11, 79, 96, 99	10, 79, 99
2	M042263	10, 11	10, 11, 70, 79, 96, 99	79, 99
3	M062237	10	10, 79, 96, 99	79, 99
4	M052087	10	10, 10, 79, 96, 99	79, 99
5	M032728	A	A, B, C, D, 6, 7, 9	B, C, D, 7, 9
6	M052105	10	10, 79, 96, 99	79, 99
7	M012022	A	A, B, C, D, E, 6, 7, 8, 9	C
8	M062095	A	A, B, C, D, 6, 9	B
9	M062074	D	A, B, C, D, 6, 9	B
10	M052092	A	A, B, C, D, 6, 9	B, C, D, 9
11	M042112	D	A, B, C, D, 6, 9	A, B, C, 9
12	M062242	A	A, B, C, D, 6, 9	B, C, D, 9
13	M042202	B	A, B, C, D, 6, 9	A, C, D, 9
14	M062241	10	10, 79, 96, 99	79, 99
15	M032545	10	10, 70, 79, 96, 97, 99	70, 79, 97, 99
16	M052090	C	A, B, C, D, 6, 9	A, B, D, 9
17	M032163	C	A, B, C, D, 6, 7, 9	A, B, D, 7, 9
18	M022042	E	A, B, C, D, E, 6, 7, 8, 9	A, B, C, D, 7, 9
19	M012046	D	A, B, C, D, 6, 7, 8, 9	A, B, C, 7, 9
20	M012029	E	A, B, C, D, E, 6, 7, 8, 9	A, B, C, D, 7, 9
21	M041125	A	A, B, C, D, 6, 9	B, C, D, 9
22	M031227	10	10, 19, 79, 96, 97, 99	79, 97, 99
23	M031251	B	A, B, C, D, 6, 7, 9	A, C, D, 7, 9
24	M041124	10	10, 70, 71, 79, 96, 99	70, 71, 79, 99
25	M061254	10	10, 70, 79, 96, 99	70, 79, 99
26	M031242A	10	10, 70, 71, 79, 96, 97, 99	79, 99
27	M031247	20	20, 10, 11, 70, 79, 96, 97, 99	70, 79, 99
28	M051006	20	20, 10, 11, 79, 96, 99	79, 99

A.3 Syntax Used to Recode Variables

To recode the variables, we used the Statistical Package for Social Sciences (SPSS) software (Version 24). SPSS is a statistical software used for data entry, cleaning, coding and performing variety of statistical analyses. The SPSS software is available for purchase on the IBM website (https://www.ibm.com/products/spss-statistics).

Below we present as an example, the recoding syntax for physics item 1A.

Example SPSS syntax for recoding variable to set "Not Reached" as a missing value

```
RECODE PA33061A (ELSE=Copy) INTO PA33061A_REC_NR.
VARIABLE LABELS PA33061A_REC_NR 'ACCELERATION OF VERTI
CALLY THROWN BALL (D) REC NR'.
EXECUTE.

APPLY DICTIONARY
/FROM *
/SOURCE VARIABLES=PA33061A
/TARGET VARIABLES=PA33061A_REC_NR
/FILEINFO
/VARINFO ALIGNMENT FORMATS LEVEL ROLE MISSING VALL
ABELS=REPLACE ATTRIBUTES=REPLACE WIDTH.

* Define Variable Properties.
*PA33061A_REC_NR.
MISSING VALUES PA33061A_REC_NR(6).
VALUE LABELS PA33061A_REC_NR
1 'A'
2 'B'
3 'C'
4 'D*'
6 'Not reached'
9 'Not Answered'.
EXECUTE.
```

Example SPSS syntax for recoding variable for scoring

```
RECODE PA33061A_REC_NR (ELSE=Copy) INTO PA33061A_REC_SCR.
VARIABLE LABELS PA33061A_REC_SCR 'ACCELERATION OF VERTI
CALLY THROWN BALL (D) REC SCR.
EXECUTE.

RECODE PA33061A_REC_SCR (1=0) (2=0) (3=0) (4=1) (9=0) (6=6)
(ELSE=SYSMIS).
```

```
EXECUTE.
APPLY DICTIONARY
/FROM *
/SOURCE VARIABLES=PA33061A_REC_NR
/TARGET VARIABLES=PA33061A_REC_SCR
/FILEINFO
/VARINFO ALIGNMENT FORMATS LEVEL ROLE ATTRIBUTES=
REPLACE WIDTH.

* Define Variable Properties.
*PA33061A_REC_SCR.
MISSING VALUES PA33061A_REC_SCR(6).
VALUE LABELS PA33061A_REC_SCR
1 'D*'
0 '0'
6 'Not reached'.
EXECUTE.
```

Example SPSS syntax for recoding variable for misconception

```
RECODE PA33061A_REC_NR (ELSE=Copy) INTO PA33061A_REC_MIS.
VARIABLE LABELS PA33061A_REC_MIS 'ACCELERATION OF VERTI
CALLY THROWN BALL (D) REC MIS'.

EXECUTE.

RECODE PA33061A_REC_MIS (1=2) (2=0) (3=0) (4=1) (9=0) (6=6)
(ELSE=SYSMIS).
EXECUTE.


APPLY DICTIONARY
/FROM *
/SOURCE VARIABLES=PA33061A_REC_NR
/TARGET VARIABLES=PA33061A_REC_MIS
/FILEINFO
/VARINFO ALIGNMENT FORMATS LEVEL ROLE ATTRIBUTES=
REPLACE WIDTH.

* Define Variable Properties.

*PA33061A_REC_SCR.
MISSING VALUES PA33061A_REC_MIS(6).
VALUE LABELS PA33061A_REC_MIS
0 'Other Incorrect'
```

```
1 'Correct'
2 'Misconception (A)'
6 'Not reached'.
EXECUTE.
```

A.4 Calculating Item Statistics

After each of the items were recoded, item statistics were calculated using the IEA's International Database (IDB) Analyzer software, using the *Percentages* function. The same types of analyses of TIMSS and TIMSS Advanced data can also be conducted using the National Center for Education Statistics' Ed*Survey* R Package.

A.4.1 More on the IDB Analyzer

The IDB analyzer (https://www.iea.nl/data) is a free application developed by the IEA that can be used to combine and analyze data from IEA's large-scale assessments. The IDB Analyzer takes into account the complex sample survey design of using multi-stage clusters and the use of plausible values required for each study. Although the IDB Analyzer is free, users still need access to either SPSS or the Statistical Analysis System (SAS) software (another statistical software) to run the syntax created by the IDB analyzer. For more information on how to use the IDB Analyzer, please refer to the *HELP Manual*, which is available once the software is downloaded.

A.4.2 More on the Ed*Survey* R package

Researchers looking for cost-effective software to analyze the international datasets can also use the Ed*Survey* package, available as open-source R software (https://cran.r-project.org/web/packages/EdSurvey/index.html), which produces similar estimates, as well as the related *t*-tests. Developed by the American Institutes for Research (AIR) and commissioned by the National Center for Education Statistics (NCES), the Ed*Survey* package gives users the ability to process and analyze large-scale international datasets efficiently, taking into account the complex sample survey design using multi-stage clusters, and the use of plausible values required for each study. A reference manual (https://cran.r-project.org/web/packages/EdSurvey/EdSurvey.pdf) is available to help inform researchers using the Ed*Survey* package, as well as several vignettes (https://cran.r-project.org/web/packages/EdSurvey/vignettes/introduction.html) that provide example analyses. Descriptions of the statistical methods and how to use the Ed*Survey* package to conduct analyses with TIMSS data can be found here:

- Statistical Methods Used in Ed*Survey*: https://www.air.org/sites/default/files/EdSurvey-Statistics.pdf
- Using Ed*Survey* to Analyze TIMSS Data: https://www.air.org/sites/default/files/EdSurvey-TIMSS.pdf (p. 25)

Users can also use the *Help* function while they are using the Ed*Survey* R Package to obtain additional details.

A.5 Calculating Gender Differences

The administrative variable ITSEX was used as a grouping variable to calculate item statistics by gender. As described in Chap. 3, analyses for gender differences were conducted two ways for select items: first using the Ed*Survey* R Package (Version 2.0.3) *Gap* function and then using the IDB Analyzer (Version 4.0) *Percentages* function. The output from both methods was compared to evaluate any impact on the statistical significance of the gender differences using the approximated standard errors (independent *t*-test) obtained using the IDB Analyzer instead of the fully correct standard errors (non-independent *t*-test) obtained using the Ed*Survey* R Package.

We here present one example item obtained using the IDB Analyzer (Table A.3) and the Ed*Survey* R Package (Table A.4), as well as the resulting calculated *t*-statistics. Percentages and standard errors for females and males obtained from the IDB Analyzer output, and the calculated gender differences, standard errors, and *t*-statistics using the independent *t*-test formula can be compared with the corresponding output obtained using the Ed*Survey* R Package, which shows the same percentages and standard errors for females and males obtained using the IDB Analyzer (Tables A.3 and A.4). The output from the Ed*Survey* R Package also includes the covariance between females and males, the female–male difference, and the standard error of the difference. The covariances between males and females were very small across all countries (e.g., ranging from –0.00039 to +0.00018% for response option A). The calculated *t*-statistics based on the non-independent *t*-test (rounded to 0.01) were not measurably different from those obtained using the IDB analyzer output. Similar results were found for other items tested using both analysis methods. Based on these results, we used the IDB Analyzer output to calculate gender differences for all items.

Table A.3 Physics item 1A (PA33061A) statistics generated using IDB Analyzer, 2015

Country	Response	IDB Analyzer output				Calculated statistics		
		Female (%)	Female (SE)	Male (%)	Male (SE)	Difference F − M (%)	Difference F − M (SE)	*t*
Italy	A	66.07	2.64	74.82	1.95	−8.75	3.28	−2.67
Norway	A	53.22	3.47	33.44	2.34	19.77	4.18	4.73
Russia	A	48.41	3.25	46.98	2.37	1.43	4.02	0.36
Slovenia	A	55.81	4.48	38.35	2.93	17.46	5.35	3.26
United States	A	50.66	4.33	34.96	4.25	15.70	6.07	2.59
Italy	B	9.99	2.19	5.58	0.94	4.41	2.38	1.85
Norway	B	2.84	1.12	1.69	0.59	1.15	1.27	0.91
Russia	B	3.98	0.95	1.96	0.63	2.01	1.14	1.77
Slovenia	B	0.00	N/A	3.61	1.40	−3.61	1.40	−2.58
United States	B	6.12	2.21	1.11	0.44	5.01	2.26	2.22
Italy	C	10.47	1.50	6.19	1.17	4.29	1.90	2.25
Norway	C	3.21	1.14	1.59	0.49	1.62	1.24	1.31
Russia	C	5.97	1.22	1.81	0.68	4.16	1.40	2.98
Slovenia	C	5.39	2.32	3.97	1.39	1.42	2.71	0.52
United States	C	6.74	2.28	1.54	0.54	5.19	2.34	2.22
Italy	D*	6.82	1.46	10.81	1.54	−3.99	2.13	−1.88
Norway	D*	39.86	3.76	63.12	2.33	−23.25	4.43	−5.25
Russia	D*	40.71	3.33	48.59	2.62	−7.89	4.23	−1.86
Slovenia	D*	37.19	4.36	52.89	3.21	−15.70	5.41	−2.90
United States	D*	35.43	4.16	60.73	3.88	−25.30	5.69	−4.45
Italy	Omitted	6.65	1.29	2.61	0.65	4.04	1.45	2.79
Norway	Omitted	0.87	0.63	0.16	0.16	0.71	0.65	1.08
Russia	Omitted	0.94	0.35	0.65	0.29	0.28	0.45	0.63
Slovenia	Omitted	1.61	1.56	1.17	0.87	0.43	1.79	0.24
United States	Omitted	1.06	0.60	1.66	1.09	−0.60	1.25	−0.48

Notes Calculated *t* values are based on the independent *t*-test. SE = standard error

Table A.4 Physics item 1A (PA33061A) statistics generated using R Ed*Survey* Package, 2015

Country	Response	R Ed*Survey* output							Calculated statistics
		Female (%)	Female (SE)	Male (%)	Male (SE)	Difference F (%) − M (%)	Difference F − M (SE)	Covariance	*t*
Italy	A	66.07	2.64	74.82	1.95	−8.75	3.28	0.00011	−2.67
Norway	A	53.22	3.47	33.44	2.34	19.77	4.18	0.00018	4.73
Russia	A	48.41	3.25	46.98	2.37	1.43	4.02	0.00018	0.36
Slovenia	A	55.81	4.48	38.35	2.93	17.46	5.35	0.00007	3.26
United States	A	50.66	4.33	34.96	4.25	15.70	6.07	−0.00039	2.59
Italy	B	9.99	2.19	5.58	0.94	4.41	2.38	0.00002	1.85
Norway	B	2.84	1.12	1.69	0.59	1.15	1.27	0.00000	0.91
Russia	B	3.98	0.95	1.96	0.63	2.01	1.14	0.00003	1.77
Slovenia	B	0.00	N/A	3.61	1.40	−3.61	N/A	N/A	N/A
United States	B	6.12	2.21	1.11	0.44	5.01	2.26	0.00000	2.22
Italy	C	10.47	1.50	6.19	1.17	4.29	1.90	0.00005	2.25
Norway	C	3.21	1.14	1.59	0.49	1.62	1.24	0.00001	1.31
Russia	C	5.97	1.22	1.81	0.68	4.16	1.40	0.00003	2.98
Slovenia	C	5.39	2.32	3.97	1.39	1.42	2.71	−0.00010	0.52
United States	C	6.74	2.28	1.54	0.54	5.19	2.34	−0.00001	2.22
Italy	D*	6.82	1.46	10.81	1.54	−3.99	2.13	0.00002	−1.88
Norway	D*	39.86	3.76	63.12	2.33	−23.25	4.43	0.00032	−5.25
Russia	D*	40.71	3.33	48.59	2.62	−7.89	4.23	0.00032	−1.86
Slovenia	D*	37.19	4.36	52.89	3.21	−15.70	5.41	−0.00012	−2.90
United States	D*	35.43	4.16	60.73	3.88	−25.30	5.69	−0.00022	−4.45
Italy	Omitted	6.65	1.29	2.61	0.65	4.04	1.45	0.00000	2.79
Norway	Omitted	0.87	0.63	0.16	0.16	0.71	0.65	0.00000	1.08
Russia	Omitted	0.94	0.35	0.65	0.29	0.28	0.45	0.00000	0.63
Slovenia	Omitted	1.61	1.56	1.17	0.87	0.43	1.79	0.00000	0.24
United States	Omitted	1.06	0.60	1.66	1.09	−0.60	1.25	−0.00002	−0.48

Notes Calculated *t* values are based on the non-independent *t*-test. SE = Standard error

References

Averett, C., Ferraro, D., Tang, J., Erberber, E., & Stearns P. (2017). *Trends in International Mathematics and Science Study (TIMSS), U.S. TIMSS 2015 and TIMSS Advanced 1995 & 2015 Technical Report and User's Guide* (NCES 2018-020). US Department of Education. Washington, DC: National Center for Education Statistics. Retrieved from https://nces.ed.gov/pubsearch/pubsinfo.asp?pubid=2018020.

The manufacturer's authorised representative in the EU is Springer
Nature Customer Service Centre GmbH, Europaplatz 3, 69115 Heidelberg,
Germany. If you have any concerns regarding our products, please
contact ProductSafety@springernature.com

Printed and bound by CPI Group (UK) Ltd, Croydon, CR0 4YY
30/09/2026
02225608-0017